NORTH CAROLINA
STATE BOARD OF COMMUNITY COLLEGES
LIBRARIES
CAPE FEAR TECHNICAL INSTITUTE

20.00

D1087716

Scallops and the Diver-Fisherman

Scallops and the Diver-Fisherman

David Hardy

Fishing News Books Ltd
Farnham · Surrey · England

Property of Library
Cape Fear Technical Institute
Wilmington, N. C.

© David Hardy 1981

British Library Cataloguing in Publication Data

Hardy, David
Scallops and the diver-fisherman.
1. Scallops
2. Scallop fisheries – Great Britain
I. Title
639′.48   SH371.5
ISBN 0 85238 114 X

Published by
Fishing News Books Ltd
1 Long Garden Walk
Farnham, Surrey
England

Set in 11/12 VIP Plantin by Inforum Ltd, Portsmouth
Printed by the Pitman Press Ltd, Bath

Contents

Illustrations

Publisher's note

The author of this book, David Hardy, is an experienced diver. He acquired a Higher National Diploma in Business Studies in 1970, but after taking a course on sub-aqua diving he soon adopted a career as a diver-fisherman. He began this career by diving for crawfish and scallops but with the development of oil exploration projects at sea, acquired the necessary additional skills to undertake commercial diving work in those fields. In addition he served as Master of a 150 ton cargo vessel serving the North Sea Ninian platform whilst it was under construction. However his enthusiasm for the role of the diver-fisherman never diminished. This enthusiasm is expressed in this book together with many practical hints on this form of fishing.

Although drawing on experiences and information from other parts of the world he writes primarily from the viewpoint of a British diver-fisherman. Readers in other countries would need to seek information from their respective government or other authorities on where the very essential skill in diving can be obtained, as it can be very dangerous without adequate training. It will also be necessary to obtain information on such regulations as may exist on sub-aqua diving, fishing, and the use of coastal waters for shell fish farming, before contemplating the activities outlined in this book.

The author refers to specific manufacturers of diving equipment obtainable in the UK. This does not preclude the availability of similar equipment in other countries but experienced advice on the suitability of such equipment for the use of a diver-fisherman should be sought locally.

Introduction

My first purpose in writing this book has been to outline scallop diving as a satisfying and remunerative activity for the diver, offering a great deal of freedom but requiring the skill and patience of a fisherman as well as diving expertise. This is important: it is quite easy to teach a man to dive; far more difficult to teach him to be a good fisherman. The scallop diver succeeds by results; he is paid for what he finds, and that too is an attraction. The competitiveness of scallop diving is another of its rewards; but if this leads to greed and hurried fishing by unsafe means, it is likely to end in tragedy. I cannot stress too strongly that individual divers must operate sensibly and refrain from dangerous cut-throat competition. The job is one to enjoy, with rewards measured by activity and skill.

A second purpose in encouraging scallop diving is that it can help to conserve an important natural resource for future harvests. The diver sees scallops in their natural surroundings and his observations can contribute towards a better understanding of their habits and life-cycle. Divers and scallop dredgers have worked side by side for many years, and will no doubt continue to do so, but it is important to recognize that the scallop dredge achieves results at some cost to young life in the beds, but a diver can harvest selectively and can even establish 'farms' on the sea-bed where small scallops or scallop larvae can be grown to marketable size.

My third reason for writing is that I am an enthusiast for the craft and skill of diving and want to share what I can of my experience. Every diver encounters his own problems and eventually arrives at the working methods that suit him best. However, I hope that all divers will derive some benefit from my description of the more successful methods of fishing scallops and other species.

I hope very much that traditional fishermen too may find something of interest here. Some devotees of orthodox methods have yet to be won over to the view that scallop diving has become an efficient and well-organized form of fishing. Perhaps their mistrust results from the activities of some holiday and week-end divers who are not commercially integrated into the industry. Good relations between divers and traditional fishermen are important. Each can help the other – but it takes only one instance of ignorant or inconsiderate behaviour to ruin good under-standing with men possessed of older and more estab-lished skills. If this book promotes further recognition by traditional fishermen of the professional status of the scallop diver it will have fulfilled another vital purpose.

1 The diver-fisherman

It can be safely said that of all types of commercial diver the diver-fisherman is the longest established. For thousands of years men have been diving to gather shell-fish from shallow pools, and the Australian aborigine has long used a reed 'snorkel' as a means of approaching wildfowl from under the water. Sponge divers, pearl divers and coral divers have used their skills for centuries, but the introduction of suitable modern diving gear allowed the practice to grow in popularity. With the radical development of diving equipment and techniques, it became feasible for a hitherto inaccessible two-thirds of the world's surface to be explored.

Diver's work

Divers today can be split up into seven groups: the high-prestige oil diver; the civil engineer diver; the salvage diver; the service diver of the army, navy, air force and police; the scientific diver (engaged in marine biology, nautical archaelogy, geology *etc*); the club diver; and the diver-fisherman. The growing popularity of diving, both as a profession and as a sport, is remarkable and has been stimulated by the increasing demand for underwater workers as well as the adventure of exploring the hitherto unknown.

The diver-fisherman is always looking out for marine

creatures that might prove a profitable diving proposition for the future. It is important that he must be able to accept that although he may be proud of his skill as a diver, his success or failure will depend mainly on his fishing ability.

The diver-fisherman may find himself alienated from the rest of the diving world. Lack of communication between types of divers has been one of the causes of this problem, coupled with the diver-fisherman's total independence from outside influences. Most forms of diving are regulated in some way, and rules have been laid down to ensure that work is carried on within tight controls. The shellfish diver has tended in the past to work outside these controls. To correct this lack of integration will require some reappraisal of attitudes all round. I once worked with a couple of divers who refused to wear life jackets. Their reason was that sub-aqua clubs made this practice compulsory, and that they had heard that sub-aqua clubs referred to shell-fish divers as 'cowboys'. They therefore refused to do anything that sub-aqua clubs laid down. This was, of course, no way to react, and it shows what can happen when ill-judged or ignorant comments are made.

The introduction of self-contained under-water breathing apparatus (scuba) enabled the diver-fisherman to explore areas which were impossible to reach in the past, and has given him freedom and mobility. While the newly found freedom of movement widened the scope of the industry, it also, unfortunately, made diving so easy that vast areas of shellfish were cleared by divers in a very short time. It was noticeable that, while the rest of the diving world quickly took advantage of technical innovations and advances, shellfish divers tended to use the most basic gear, in the hope that improvements in practice would emerge from basic fishing skills.

Training

It is a good practice for a beginner to start as a member of a crew with payment by results – the object being development of competition and incentive. Experience with beginners shows that when they see an experienced diver working nearby topping their catch their complaint is that they have been put on poor ground. 'After all', they are apt to say 'it is only like picking up potatoes in a field, isn't it?' In fact, it is not at all like that. Potatoes are planted evenly in straight rows; scallops go where they like, and it takes a great deal of experience and observation of underwater surfaces to acquire an understanding of the sites and environment that scallops prefer.

Dangers and difficulties

It should be remembered by the beginner that a diver enters a strange environment when he goes underwater and this can affect him physically and mentally. He is restricted in many ways when on the sea-bottom, so he must be in good mental and physical shape before he even thinks of venturing down. Each day the shellfish diver goes to work he has to cope with such dangers and difficulties as nitrogen narcosis; breathing compressed air; distortion of vision and hearing; restricted movement; the anxieties of a strange environment and of having to find a suitable fishing ground; cold; physical exhaustion and pressure.

Nitrogen narcosis

Nitrogen narcosis is something that begins to affect the diver from a depth of about 90 feet downwards. It is a result of a build up of excess nitrogen in the body and affects the diver in a very strange way. When it comes on he will begin to feel intoxicated and his powers of rational thinking will gradually be weakened; he may lose up to a third of his reasoning ability and the length of time he takes to react may be increased by one fifth. This, of course, will not help any planning he may have to do on the bottom concerning his fishing. The only way to overcome nitrogen narcosis is to breathe a mixed gas with a higher level of oxygen and a lower level of nitrogen, but this is much too expensive for the scallop diver.

Breathing

Many problems arise in supplying a man in the water with air. When air is compressed much care is needed to ensure that it does not become contaminated with oil, water, or other impurities. If the oil and water are not separated properly the air will be left with a distinct taste, which makes it unpleasant to breathe and which can be dangerous to the diver's health. If carbon monoxide is drawn into the compressor it will have a very adverse effect on the diver, and in large doses could prove fatal. Unfortunately, although the compressor's charcoal filter will extract most impurities from the air it will not absorb carbon monoxide.

Carbon dioxide can also create a problem for the diver if breathed in excess. The problem lies in expelling the diver's exhaust gases, which contain a high level of carbon dioxide. Full face masks are particularly likely to allow large amounts of this gas to build up, and the mast must be flushed through regularly with air.

Breathing also demands slightly more effort than usual because the demand valve is opened by reduction in pressure; the diver has to suck on his valve to get air. Modern demand valves are very sensitive and respond to the smallest pressure change, but it is still not as easy to breathe as on the surface. 'Free-flow' systems that help to overcome this problem, are hardly suitable for the shellfish diver because they would restrict his movement and use up too much air.

Distorted sight and hearing

Objects appear to be approximately one third larger and slightly refracted when viewed under water. Add to this the slight distortion caused by the mask's face-plate, and it will be realized that the diver has a difficult job to estimate the size and distance of objects on the bottom. Light passing through the water loses its intensity quickly, being readily reflected from any particle which gets in its way. Light is also distorted when it meets currents of water with different temperatures, and if there is a layer of fresh water on the surface, this too can alter the characteristics of the light. All this makes focusing very difficult, and as the diver's hearing is also affected underwater, he has to work with his two most useful senses impaired.

Restricted movement

In most places the water is not warm enough to dive without wearing some type of insulation, and the diver must work in a suit which, although it keeps him reasonably warm, restricts his movement. The water itself also resists movement and, although he works in a kind of slow motion, a great deal of the diver's energy is used in overcoming that resistance.

Anxiety

The strangeness of his underwater environment has an effect on the diver, and most suffer from some kind of anxiety while on the bottom. The risks involved when working in shark-infested or polluted water, strong currents and poor visibility are not the cause of the anxiety; it is the fear of them in the diver's mind that is the problem. The diver may have a phobia, such as a fear of dark, enclosed spaces, and it has even been known for men to lose their nerve in the presence of moving seaweed on the sea bed. Divers will often not admit to these anxieties, and consequently get no advice or reassurance when entering the water.

The shellfish diver is faced with an additional worry: the problem of not being able to find a suitable fishing ground. If he does not fill his catching bag his wage packet suffers, because he only gets paid for what he catches. Coupled with this is the anxiety of falling too far behind the other divers from the boat, so that he feels he must push himself all the time. This in turn leads to worries about whether or not he is within his body's safe working limits. Perhaps it could be said that competition over catches among divers is not a good policy, but when worked sensibly it does not lead to trouble.

7

Effect of cold

Working in cold water will affect a diver's performance and prolonged exposure will have a very marked effect on his work rate and may lead to the dangerous lowering of body temperature known as hypothermia. The diver's body dissipates heat all the time he is in the water, and this leads to a gradual deadening of the senses and a slowing-down of reactions. It soon becomes apparent after a dive that cold has had an effect, the response being drowsiness and a general slowing-down of actions. Cold can also affect the eyesight and the urinatory system, and is one of the most serious limiting factors with which the diver has to contend.

There have been descriptions of 'progressive symptomless hypothermia' leading to unconsciousness and death without warning signs. This is more likely to occur in thin individuals and is clearly a very great danger.

Pressure

Changes in water pressure are the cause of many diver's ailments, ranging from a ruptured ear drum to a burst lung. The danger lies in the compression of air trapped in the body during the descent and consequent pressure increase, and its expansion during the ascent. For instance, a balloon full of air taken to a depth of 33 feet will halve in volume. If at this depth it were to be blown up to its original size, and then taken back up to the surface, it would be found to be twice its initial size.

Ear barotrauma (burst ear drum) can be caused by a blockage in the outer ear or by congestion in the eustachian tubes. Either air will be prevented from passing into the ear during descent, or prevented from escaping during ascent.

Dental barotraumas become apparent on the ascent, and are caused by air bubbles being unable to escape from cavities in the teeth.

The lungs are particularly susceptible to damage by pressure changes: the results usually become apparent after the ascent. There are three ways a diver may damage his lungs and all are caused by not releasing built-up air pressure whilst surfacing. The first is an air embolism ('burst lung'), thought to be caused by the escape of air into damaged pulmonary blood vessels when the lung is subject to over-stretching: the bubbles of air may then pass into the heart and obstruct circulation to the brain, causing strokes. Treatment is immediate re-compression in a chamber. The second type of damage is a lung collapse and is called a pneumothorax. It is caused by air forcing its way into the lung cavity and actually squeezing the lungs from the outside. This too requires immediate re-compression. Acute surgical emphysema is the third type of lung damage, and is caused by air finding its way into the surrounding tissue. Although not as serious as the previous two, it still needs to be carefully looked after. In this case the air will actually rise through the tissue and will give off a crackling sound when disturbed.

Decompression sickness is possibly the biggest danger to the diver and results can be fatal. Shellfish divers are often accused of neglecting proper decompression procedures, but on the whole most work well within the limits of the decompression tables. Some people have a very high resistance to decompression sickness and it is quite amazing what they can tolerate, but these rare individuals should not be used as a basis for water exposure times because they are just not comparable to the average diver's endurance.

Before the problems of decompression can be fully understood one must first know something about the physiology of the body and the nature of the surrounding environment. The air we breathe is composed of approximately 80% nitrogen and 20% oxygen. When we breathe in, air enters the lungs and is transferred to the blood system by a chemical reaction in the red corpuscles. The blood stream then distributes oxygen to the body's tissues, exchanging it for used gases which are taken back to the lungs to be exhaled. This system is in balance with the body's tissues, and maintains the same pressure as the surrounding atmosphere.

Maintaining the right proportion of gases is important to this whole system and is governed by the 'partial pressure' principle. Partial pressure is quite simply the proportion of a particular gas in a composition of gases. For example, in air we have a composition of oxygen and nitrogen, and if oxygen is 20% of the composition, then its partial pressure is said to be one fifth, or 0·20. This proportion remains constant under pressure, a fact which must be borne in mind if the effect of pressure on the body is to be understood.

When the body is functioning at normal atmospheric pressure, gaseous exchanges take place within it due to partial pressures trying to stay in balance. For instance, as the body's cells produce carbon dioxide, the partial pressure of the air in the cells is altered and they readily accept oxygen from the blood stream to get back in balance. In this way oxygen is distributed to all the body's cells.

When a diver enters the water he is supplied with air at the same pressure as the head of water above him, and this presents his body with a problem. At a depth of 33 feet his body is subjected to twice the atmospheric pressure at the surface. He now needs double the pressure of air to maintain the same volume in his lungs (Boyle's Law). His body's cells now want to operate at the new pressure level and as depth and pressure increase, so his cells will absorb more gas. Active tissues such as muscles will absorb gases faster than inactive tissues, such as fat and cartilage, because there is a better blood supply to the former. The process of balance within his body runs smoothly while the pressure increases, but when pressure is reduced on ascent, problems can arise. Air trapped in his body will expand due to the reduction in pressure. The main air cavities; the lungs, the middle ear, and the sinuses, are all able to expel this extra volume of air quickly and easily but the body's tissues are not, and they take longer to dispel the gas build-up. Therefore, once a diver is back on the surface after working under pressure, his tissues take some time to get back into balance, and those that took the longest to reach their pressure level at depth will take the longest to release the gas.

If the diver lets his tissues soak up too much gas by staying under pressure for longer than he should, then when he releases the pressure, the gas, instead of staying in solution, will expand and form tiny bubbles. These bubbles travel around the body and cause the damage that is known as decompression sickness. An illustration of what happens in a diver's body when he surfaces too rapidly can be seen when unscrewing the top from a bottle of fizzy lemonade quickly. Tiny bubbles will be seen to form as the pressure is released, but if the top is screwed off slowly then these bubbles will not appear. The diver must therefore regulate his ascent (usually to the speed of his smallest exhaled air bubbles), and if his body has absorbed too much gas then he must take much longer to surface, usually by stopping for a specific time at a specified depth.

Decompression sickness (popularly known as 'the bends' because of the odd posture the diver may adopt when the symptoms develop) may come on within minutes or hours after reaching the surface, and may take various forms.

Commonly the diver may experience pains in the shoulders and elbows or hips and knees, numbness and pins and needles in the limbs, which may lead to partial or complete paralysis. Dizziness, affected vision and vomiting may occur whilst some individuals notice bruising of muscles and itching skin rashes. A serious complication is choking and chest pain ('the chokes'). Immediate recompression is necessary to avoid death or paralysis. This can be done either in a pressure chamber or back in the water, depending on the severity of the symptoms.

Avascular bone necrosis has been responsible for putting many divers out of work. Its cause is still not fully understood and new theories are being put forward all the time. However, the most probable cause seems to be a build-up of nitrogen bubbles. The release of nitrogen in certain parts of the body after a diver's ascent can obstruct blood flow, which will lead to areas of tissue dying. This obstruction is more apparent in areas where the flow of

11

blood is small, as in the fatty deposits around long bones. Lack of blood will cause the affected bone to die. New bone will grow over its surface, stimulated by the flow of blood from other parts of the body but, if the blood flow to this new bone is also stopped, then the dead bone will eventually collapse. The hips, knees and shoulders are the most vulnerable parts of the body for bone necrosis, and the diver should have these X-rayed frequently (long bone X-rays) to ensure they are in a healthy condition.

Summary

A diver-fisherman cannot class himself as competent until he can contribute information as to the habits and the likely whereabouts of the fish he is after. He must therefore keep an accurate mental record of all the ground he has fished on, its characteristics, and how successful he was there. If possible, he must watch the activities of other divers in the area, and keep a note of what and how well they are doing. This knowledge is invaluable and will pay dividends, especially if activities are expanded to include three or four divers. There is nothing more disheartening than spending day after day searching unproductive ground, and this has forced many to give up diving. If a thorough understanding of the environment and habits of the fish is acquired the time spent in searching can be reduced.

The diver-fisherman earns his living in an alien and potentially hazardous environment. He can minimize danger by understanding and being constantly alert to the dangers that may threaten underwater, and by keeping physically and mentally fit. No-one should consider working underwater unless he knows that he is fit, and yearly medical examinations are essential.

2 Scallops

Being a bivalve, the scallop is categorized as one of the lower forms of sea creature. However, along with its close relative, the oyster, it is greatly sought after as a gastronomic delicacy.

Most people involved in fishing for scallops know little of their habits and behaviour, and are only concerned with locating the areas where scallops are abundant. It is the object of this chapter to give some insight into the scallop as a living organism, in the hope that it will stimulate further investigation and, hopefully, result in bigger catches.

The scallop belongs to a family of shellfish known as Pectinidae, and the northeast Atlantic variety is known by its Latin name, *Pecten maximus*. Its ribbed, fan-like shell is brownish white in colour, growing in some cases to a diameter of six or seven inches, and it can be found lying in depressions in the sea bed with its flat side upwards.

Spawning

In the spring and autumn the scallops spawn. They are hermaphrodites, carrying both male and female reproductive organs in the same body. The roe contains the orange-red female ovary and the creamy-white testis, but reproduction is a very uncertain process, and many of the eggs remain unfertilized. The larvae that survive attach themselves by sticky threads to seaweed or loose debris, and remain in this position until large enough to seek their place on the sea-bed.

Spawning times can vary from area to area, which is something the scallop fisherman must take into account, especially if selling the catch on a meat-weight basis. For instance, in some areas scallops may spawn heavily in spring and lightly in autumn, whilst in neighbouring areas the opposite may be true. If the fisherman is able to concentrate on areas where the scallops have not spawned heavily, he can capitalize on their extra meat weight.

Size and quality

Scallops feed on microscopic organisms and the richness of the water has a distinct bearing on their quality and density. For instance, those scallops found close to the shore are noticeably larger than their deep-water neighbours of the same age, because of the higher concentration of rich feeding found in shallow water. The type of bottom also has a bearing on the quality of the scallops because some types are richer organically than others, and these will naturally sustain higher densities of scallops, usually of a greater size. Although a scallop four to five years of age is of fair marketable standard, it is not uncommon to find scallops in excess of ten and twelve years old. Like other kinds of shellfish, older ones yield a meat of poorer quality than younger ones, although in the case of scallops there is no price penalty imposed, as there is with lobsters. In fact, most buyers will give a much higher price for large scallops because of the greater meat yield.

Predators

The scallop has a number of enemies apart from divers and dredgers, and these predators can soon destroy large areas of scallops. Apart from the tightness with which it can close its shell, the only defence the scallop has against aggressors is to move, by jetting water from either side of the hinge, (which gives an appearance of biting at the water (*Fig 1*). Star fish, dog whelks and crabs seem to be the scallop's most formidable enemies, and they all indulge in devious methods of attack. The starfish will site itself right over the top of the scallop and clamp the shell closed by working its arms underneath (*Fig 2*). This position is held until the scallop is exhausted, and gives up the fight. The dog whelk, on the other hand, just waits patiently beside the shell until it can get a foot-hold inside it.

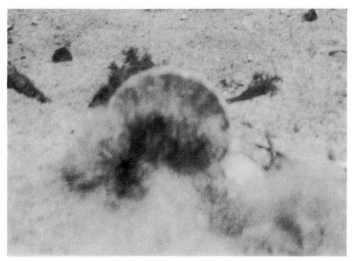

Fig 1 A scallop moving on the sea-bed by ejecting jets of water at either side of the hinge of its shell.

Small crabs are very interesting predators to watch because some species will station themselves beside the scallop for hours on end, flicking grains of sand into the shell when it is partly open. After a while the scallop will be so irritated that it will open up completely, and then the crab will insert its claw and sever the muscle (*Fig 3*).

In general there is a good balance between scallops and their predators, and the latter rarely seem to gain the upper hand. It has been interesting to note that scallops were never really overwhelmed even when thick beds were thinned out. In these instances the only predator that

15

Fig 2 A starfish positions itself over a scallop. Its method of attack is to hold the shell closed until the scallop is exhausted.

Fig 3 A small crab disturbed at its vigil beside a scallop.

seemed to cause any trouble at all was the brittle starfish and that, on the whole, was not much of a threat. It was once thought that the ecology of the sea-bed would change once the vast concentrations of scallops were thinned out, but this has proved not to be so. In fact it has become increasingly obvious that the more frequently a piece of ground is fished, the quicker the scallops come back to it. This, of course, does not apply to all areas; there are always exceptions to a rule, but most ground, if fished sensibly, will begin to be restocked almost immediately. Regular fishing of beds has the advantage of thinning-out the shellfish and thereby lessening the likelihood of disease spreading through them. Vast areas of dead scallops can often be found on the bottom and a likely explanation is that the scallops bred so prolifically that when disease struck it spread very quickly.

Scallop fisheries

There is a thin scattering of scallops all around the British Isles, although some areas support much denser concentrations than others. In general, the main areas for commercial fishing are the English Channel, the area around the Isle of Man and Northern Ireland, and the west coast of Scotland. In all these areas scallops are found in sufficiently dense patches to warrant either dredging or diving for them. Areas that have long flat expanses of sand are, naturally, more suitable for dredging, whilst those with rocky outcrops or glacial deposits make good diving ground. The dredges and the divers can therefore work side by side in reasonable harmony. Few areas have the combination of high organic water content, suitable bottom, suitable tide flow, shelter from heavy seas, and a suitable depth that is necessary to sustain dense beds of scallops.

Methods of fishing

Until the late 1960s, the main method of fishing for scallops was by dredge (*Fig 4*). Then divers began to fish

17

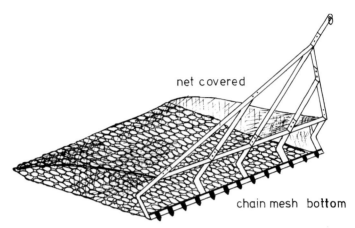

net covered

chain mesh bottom

Fig 4 A scallop dredge. The main framework, teeth, and the bottom of the bag, are metal. The top of the bag is of netting.

for them, and diving became more and more popular as the price per scallop rose. Diving introduced a new dimension to scallop fishing and the two forms, dredging and diving, seemed to work well together. As a general rule, the best beds of scallops are found at depths of between 10 and 20 fathoms, and whilst the diver can work the shallower beds the dredge can work more economically at greater depths. The diver benefits by this arrangement because scallops always seem to move from deeper into shallower water, if at all possible, so ground that has been cleared by diving should always be in the process of replenishment.

One other method of scallop fishing used in the past was with a glass box and a pole. This was only practised in water of less than 25 feet in depth, and was an extremely tiring job. The glass box enabled the scallops to be seen from the surface if the water was not too dirty, rough or deep. A long weighted pole was then used to aim a hard-eyed net bag at the scallops to gather them up. This was a very difficult job in rough weather, or in areas of strong tides, and it was only the most resolute of men that practised it. It was, however, an adequate method of obtaining a few scallops for personal consumption.

18

Markets

Most scallops landed in Britain are processed and exported either to the Continent or to America. It is this export potential that boosts the price because the consumers there are willing to pay more for their delicacies. The British are not great shellfish eaters, so the local market is not very big. There have, of course, been ups and downs in the scallop market over the years but it has kept reasonably stable, only shortages in landings and increases in demand tending to inflate the price. Decreases in demand have rarely been the cause of decreases in price; these have usually been the direct result of adverse exchange rates between Contintental and American currencies and the pound. Unfortunately this is something that the fisherman cannot control.

Processing

When the scallop arrives live at the processors, the meat is cut out of the shell (shucked), and then put into tanks of water to compensate for the dehydration that has taken place during storage on the boat and during transport (*Figs 5, 6*). A polyphosphate solution is often used to reduce dehydration to a minimum. Recommended soaking times are usually from thirty minutes to one hour, but this time is greatly increased in many processing plants. After this the meat is blast frozen and then dipped into water again for glazing. It is then transferred to the deep freeze for storage. It is interesting to note that a soaking time of thirty minutes in a polyphosphate solution and a subsequent glazing will increase the meat weight by over 8%. However, upon thawing the weight increase is only just over 3%. The meat in some of the older scallops is often discoloured and therefore less attractive to the consumer. Chemicals can be added to the soaking bath to bleach the meat and so present a more pleasing creamy-white appearance. It must be remembered, though, that all this is done at the expense of flavour, so it should be kept to a minimum. Prolonged cold storage can adversely affect the flavour of the meat.

Fig 5 Scallop meat being removed from the shell at a processing plant.

Fig 6 Scallop meat being soaked at a processing plant to correct dehydration that has taken place during transport and storage.

20

Processors try to obtain the highest meat yield possible, and, as a rough guide, the meat content of a scallop weighs from 10% to 15% of its total weight. The meat consists of 73% to 79% water, 19.5% protein, and 0.5% to 1.0% fat. Along with its distinctive flavour, it is this high protein and low fat content that makes the scallop such a popular shellfish.

3 Locating and picking up scallops

Where to look for scallops

Scallop divers agree that there are rules about the location of scallop fishing grounds, but the finer points of practice always seem to arouse controversy. Divers are constantly surprised to find scallops in areas that theoretically should be barren and disappointed to find none in areas that should be perfect for scallops. The lesson is not to overlook any type of ground if it seems remotely possible that there could be scallops on it. Luck obviously plays an important part in finding scallop beds, but theoretical and practical knowledge of scallops and their environment will pay dividends in the long run.

When searching for scallops, it has often been my experience to miss quite good beds by literally a couple of yards. Thus a second, and even a third, look at unfruitful ground will often give a return. Just how much time to spend looking at barren ground is a difficult decision to make. With a knowledge of where the scallops are likely to be, the most probable places can be checked quickly, leaving time to look at a few of the less obvious spots. Of course, if there is plenty of good fertile ground at hand, then less time will be spent researching unprofitable areas.

I have seen experienced divers search as much as a mile of shore at one hundred yard intervals and not come up with a worthwhile quantity of scallops. Someone else has then gone on to the same ground and taken quite a good catch from it. This illustrates that you should not take anyone's word for what a particular piece of ground is like; take a look yourself.

Charts play an important part in helping to locate good scallop ground. The depth of water and type of bottom are given on the chart, and this gives one a good point from which to start. If a graphic echo-sounder is used together with a chart, these are sufficient aids with which to start a search, provided one has a working knowledge of the scallops' preferred environment. The following factors should be kept in mind in any search: tidal movement, type of bottom, exposure of bottom and depth.

Tidal movement

Scallops are usually found in areas where there is a good tidal movement, due to the plentiful feeding provided by the greater flow of water. On the other hand, in areas with exceptionally strong tides scallop density decreases. The tide also helps scallops to move along the bottom, and it is therefore possible that scallops will be concentrated at the edge or extreme limit of a strong tide. This is a long shot but well worth looking into; periodically it pays dividends.

Although tidal movement is an important factor in scallop settlement, do not overlook areas where there is little or no tide, because they too sometimes have a thin scattering of scallops. Sometimes the tide can be flowing one way on the surface and the opposite way on the bottom; and in some areas there seems to be no tide on the surface yet on the bottom it is flowing quite swiftly. The tides must therefore be studied carefully to make sure that their full effect is understood.

Sea-bed

The growth rate of the scallop shell depends on the type of bottom it lies on, and it soon becomes obvious that certain types of bottom produce superior quality scallops. It can

23

be assumed that the scallop prefers ground where its shell will grow. In an area rich in scallops, they are to be found on almost all types of bottom. Scallops are sometimes found even on solid rock, but in these cases they have probably been washed there. What the scallop diver needs to know is which type of ground is likely to support the greatest number of scallops. In order to analyse this density I have divided the sea-bottom into the following ten types: shell sand, coarse sand, fine sand, mud, coral, gravel, sand/mud, sand/coral, sand/rock and solid rock.

Shell sand is probably the most productive type of ground, and when thick patches of scallops are found they are usually very extensive. Scallops on this type of ground are usually quite large, easily seen and therefore easily fished.

Coarse sand is not as productive as shell sand but quite good. In my experience the darker the ground the better the scallops, and once again they are easily seen.

Fine sand bottom is very inconsistent. It can sometimes maintain a thin scattering of scallops and any scallops on this ground are very easily seen.

Mud is not to be confused with silt: some scallop divers will insist they have never found anything worthwhile on a mud bottom, but I think they have made this mistake. Mud bottoms can give a good fishing but tend to be patchy; plentiful areas are separated by very barren stretches. Scallops found on mud are camouflaged by the large amount of organic growth on them, such as weed and barnacles. Fishing can also be made difficult by the darkness of the bottom and the stirred-up sediment. Tidal movement in muddy areas is sometimes not very great, and when sediment is stirred up it takes a long time to settle.

Coral is probably one of the most picturesque types of bottom, but very unproductive. Divers often talk about coral being good ground, but they are usually referring to a mixture of coral and sand or coral and mud. Scallops on this type of bottom are quite easily seen. The only problem is the abrasiveness of the coral on the diving suit; knees tend to get worn out quickly.

Gravel is not always productive but a fishing can sometimes be taken from this type of bottom if in a good tidal area.

Sand/mud is quite a productive type of bottom and it tends to be more consistent than mud alone. The scallops are easily spotted, and as the texture of the bottom gets coarser the fishing usually improves.

Sand/coral is a very productive bottom. The scallops are easily spotted, but working this ground is hard on the suit.

Sand/rock bottom, if the rock is well broken and scattered, can be quite good. The scallops are easily seen because it is difficult for them to lie flat on the bottom. Again, this type of ground is very hard on the suit.

Solid rock is unproductive, and if scallops are found on solid rock they are usually in transit to some other type of bottom. Do not ignore rock, as it must end somewhere, and its edge will usually give a good fishing.

Under most of the headings above I have mentioned the degree of ease or difficulty with which the scallops are spotted. On some types of ground they cannot be missed, whilst on others they are very difficult to spot (*Figs 7, 8, 9*). It is good policy to stop every now and then to have a good look round, especially on ground where the scallops are well hidden. It is surprising how many are missed due to going over the ground too quickly, and this periodic stopping will lead to improved catches. Many times I have stopped in my tracks and, on taking a close look around, have seen two or three shells lying beneath me. To be able to spot scallops easily is basic to scallop fishing. A knowledge of all types of growth on the bottom is

Fig 7 A diver's dream: easily-spotted scallops on a clean sand bottom.

Fig 8 A scallop on the sea-bed at a depth of 50 feet.

Fig 9 Showing how well-concealed and difficult to spot a scallop can be, even on a weed-free bottom.

essential, as this can be relevant to scallop growth and density.

Gullies are very popular and will always sustain a certain amount of re-settlement, whilst the lower edges of rock promontories usually remain very barren. The scallop could, of course, be choosing the sheltered gully because it is a more suitable place in which to reproduce. There is less likelihood of the tide or swell sweeping away the new eggs before they are properly fertilized, and it is also more likely that the scallops will be packed closer together in this type of spot, again making fertilization more certain. Another point in favour of gullies is that they are usually fertile spots. There the texture of the bottom is generally much richer than in surrounding areas, and a lot of marine growth is at hand, a combination which would encourage scallop growth.

Shelter and depth

The exposure of an area is very important when looking for scallop ground. It used to be believed that no scallops would be found in any area exposed to a heavy swell, but

27

this has since been proved incorrect. The idea evolved originally because when scallop diving first started it was restricted to shallow water, and exposed shallow water does not carry many scallops. However, scallops often lie in exposed deep water. When beds are found in exposed deep water the scallops are usually very dense but of poorer quality than those in shallower beds.

When looking at a new area remember that sheltered water is a likely scallop ground if all other factors are favourable. However, sheltered ground is not always obvious: it does not necessarily mean the inner side of an island, sheltered from the swell and prevailing wind; it can also mean a twelve fathom sand bank sheltered by a rock edge and this, of course, can only be detected with an echo-sounder.

It is difficult to give a general rule as to which depth gives the highest density of scallops; it is a factor to be considered in conjunction with the others already mentioned. If the shallow water is too exposed, then quite often the scallops will lie in deeper water where they will be given protection either by a rock bank or simply by their depth. Ten to twenty fathoms seems to be the depth at which scallops are at their most dense.

The question of depth raises a basic economic point in scallop diving. Deep water scallops, although being of poorer quality than those of shallow water, are usually more dense.

Thus, if selling on a meat weight basis, it may be advantageous to seek the larger, shallow water scallops. On the other hand, this type of fishing involves extra work as fewer scallops can be collected in a bag, and as time on the bottom is prolonged a longer day is required. The diver has also more swimming to do in shallow water.

Scallop diving tends to move from deep water to shallow water and back again as times goes by. Once all the shallow water in a particular area has been covered the divers move into the deeper water. When this becomes unproductive it is usually found that the shallow water has filled up again, and so the cycle restarts. This cycle covers a period of about eighteen months. Deep water diving is more popular in the winter because less time needs to be spent on the bottom which prevents the cold getting too much of a hold on the diver.

With a good knowledge of the effects on scallops of depth, exposure, type of bottom and tidal movement, suitable fishing ground should be found. What is more to the point is that the diver now knows the ground least likely to bear scallops. This knowledge can now be coupled with some theories on scallop re-settlement to give a more complete picture of the habits of this shellfish.

Repopulation of scallop beds

Those fishermen who have been occupied with dredging for scallops have definite views about their habits. However, one or two of their theories can be called into question by actually going down and watching the scallops on the sea-bottom. I once heard an argument that the scallop actually lies beneath the sand. If that was true, it would account very simply for the repopulation of scallop beds. The basis for this argument was that those involved in dredging found that as they towed their gear over certain ground again and again, the fishing became better and better. By the time the dredge had been over a piece of ground three or four times, landings were at a peak. The dredge fisherman had also noticed that a lot of the scallop shells were pure white, suggesting that they had come up from beneath the sand. They had concluded from all this that as the dredge digs deeper into the bottom so it brings up more scallops.

But there is another quite feasible explanation. Most of the areas a scallop diver searches are enclosed or sheltered either by rock or by kelp, so the bottom is protected, to a certain extent, from the action of the waves. However, on certain types of bottom, if the diver were to swim well out away from the edge and on to the sand he would find that it is very undulating, like the uniform ripples on a pool into which a stone has been thrown. Although there are a certain number of scallops lying on top of these waves of sand, the majority tend to lie in the troughs. When a dredge runs over this ground it picks up the scallops on the crests of the sand ridges first, and then, when it has removed the ridges altogether, it can get at those lying in the troughs.

This type of ridge formation is very common in a coral/sand type of bottom, and it can be assumed that it would take three or four runs of the dredge before enough would be removed to get at those shells at the bottom of the troughs. With regard to the whiteness of the shell, this is not uncommon particularly in coral areas. My explanation does not completely disprove the theory that the scallops lie beneath the sand, but it does help to explain why in some cases it takes three or four drags of the dredge to get profitable landings. A sand and stone bottom would give similar results, with the dredge having to move the stones out of the way before it can get at the scallops. A further argument is that the dredge, whilst moving over the ground, pushes many shells into the sand, so leaving them to be dug out on subsequent runs.

The theory that scallops develop beneath the sand also fails to explain why, in some thoroughly-fished areas, the scallops never return.

Scallops can be observed to move when a particularly noisy boat passes overhead and it has been suggested that this might account for the relatively fast re-population of some scallop beds in certain areas where there is a lot of shipping. The suggestion is based on very little evidence and it could equally well be that the scallops are moving away from the disturbance. All that can be said for sure is that the scallops are agitated by passing vessels.

By far the most common theory is that beds are restocked by the progression of scallops from deeper water to shallow water. It would appear that the scallops work their way up a slope, as opposed to down, because in general they lie facing that way. (This is why they are more easily spotted when a diver works his way down a slope.) Whether or not the scallops continue this progression into shallow water once they have found a suitable edge or gully to lie against is open to question.

The progression of scallops up a slope into shallower

water seems to make sense, but if no scallops can be found in deeper water and the ground is still being re-stocked, where do they come from? It could be that we are overlooking the fact that this re-stocking usually takes up to a year or more, so the number of scallops making their way up the slope at any one time would be very small, perhaps even so small as to be totally overlooked.

If we combine these ideas of the scallops' movement to shallow water with the possible effects of tide perhaps a more satisfactory explanation of re-population will be found. The tide can be seen to have an influence on where scallops lie and in areas where it is very strong, the scallop can be carried a good distance once it starts to swim. This sometimes leads to a build-up at a rock edge, or a distinct line of scallops where the tide forms an eddy. Quite simply then, the tide probably spreads the scallops out at all depths whilst they continue their progression up the slope into shallow water. This would at least help to explain how re-stocking can take place where there is no evidence of scallops in deeper water.

There are many places on the sea bed where warm currents meet cold currents and form a distinct ridge. This type of temperature change seems to stimulate marine growth. Could this be another factor in helping to explain the re-stocking problem, or is it possibly not widespread enough to have any great effect?

Most of nature's creatures are stimulated by changes either in atmospheric pressure or in temperature, so it would be natural to assume that scallops are affected by temperature changes on the bottom.

If a power cable is found on the bottom, it will be noticed that the marine population around it is greater than in surrounding areas, and scallops can quite often be found lying up against the cable itself. If there is power flowing through such a cable then one might assume that it would be generating a small amount of heat. This may only be measured in fractions of a degree, but possibly enough to stimulate marine life.

Although I have never witnessed it, some divers have reported seeing beds of 'queenies' move in a migratory fashion along the bottom. Is this also true of scallops? I know of no scientific evidence at all to support this theory and have never seen such a migration, but it certainly cannot be ruled out. Most shellfish move around on the bottom during the course of their lives and some, crabs for instance, travel very great distances. When watching crabs on the bottom they seem most often to be stationary, yet it is not uncommon for them to travel some hundreds of miles during their lifetime. They may become motionless as soon as they see the diver, so, although the diver may not see vast beds of scallops moving along the bottom, that does not mean that it could not happen. It would be a scallop diver's dream to witness a mass migration of scallops.

Obviously the scallop does have a purpose in its movements, because some spots become well stocked and are quickly re-populated after fishing, while other places are avoided. A stretch of shoreline can be fished very thoroughly and the scallops will then return to just the same spots that were fished before, and not to the places in between, however promising they may look.

None of the ideas examined in this chapter satisfactorily explain why the diver sees so few small scallops, or why scallops avoid some areas which seem perfect for their settlement and perhaps some aspects of the scallop's behaviour will never be understood, but it is certain that the more the diver-fisherman observes scallops, arriving at and testing his own theories, the better his fishing.

Picking up scallops

Once fishing ground is found, the problem arises of how to spot and pick up the scallops. Because a diver may land between three and four hundred scallops a day, people think that scallops are lying densely on the bottom. Novice divers may complain that they are being dropped on bad spots because they do not see scallops in thick patches. After a few months they learn that most of the time spent on the bottom is taken up in looking for individual scallops, dense patches being a bonus. If on average a diver spends one and a half hours on the bottom a day for three hundred and fifty scallops, this represents one every fifteen seconds. Think of the ground covered in these fifteen seconds, and then try to visualize how dense the scallops are. Of course, it is not quite as clear-cut as that, but it illustrates that the scallops are not just lying around for easy pickings. On the other hand consider the effort involved in picking up twelve hundred scallops in the same amount of time. The point I am trying to make is that scallop diving is a highly sophisticated form of fishing, requiring more than just the basic skills of diving, and as such takes a long time to master. Perseverance will give results in the long run.

Good scallop spotting comes after a knowledge of the different types of bottom has been acquired. As mentioned earlier, on some types of bottom they are easy to see, while on others they are more difficult. Once a scallop ground has been found, the next decision to be made is where to begin in order to get the best fishing. To start with, an edge must be found, firstly to give some location and direction, and secondly, because this is where scallops usually congregate. An edge does not necessarily mean rock meeting sand, mud, *etc*, although this is the type of edge that divers usually look for, and as often as not it gives good fishing. It remains the most popular because it is the most easily identified on the echo sounder. However, there are many other types of edge worth looking for. Any distinctive line, such as a line of kelp bordering the sand, can be termed an edge. Coral meeting sand is a good example of an edge and stands out particularly well when found. Sometimes it is difficult to follow an edge because of its lack of definition; but it becomes easier with practice.

When following a definite rock edge it soon becomes apparent that the best patches of scallops lie in enclosed areas such as gullies. This also applies to kelp edges, although the opening to the gully is sometimes well disguised. It will be noticed when working gullies that the scallops decrease in density around the lower outcrop before the gully. This is more evidence that scallops like shelter.

33

When working an edge it is best to zig-zag along it, to a distance of twelve to fifteen feet. This will make sure that the ground is worked thoroughly, and will leave a few scallops to eventually work their way back up to the edge. The tide can be of some assistance when fishing. If it is not too strong it is best to swim into it, as any sediment stirred up will be swept in the opposite direction to that of the diver. If the tide is quite strong, then a lot of effort will be wasted by swimming into it, so the diver swims with it but the sediment is swept ahead of him. In these circumstances, a tight zig-zag will give enough time for any sediment to be carried on. Working the tide efficiently can be rewarding, especially if a thick patch of scallops is found (*Fig 10*). The tendency would be to get down to work and clear out as many scallops in as short a time as possible. The experienced fisherman, however, would give a few moment's thought to the situation, and make sure that the tide was clearing the sediment stirred up. In this way few shells would be missed. If two or three hundred scallops are found lying in a small gully, the amount of sediment that will be disturbed will be considerable, thus it is essential to plan ahead.

Thick patches of scallops are the exception and not the rule, so we come back to the basic problem of spotting. Eyesight is very important, but I assume that those seeking an occupation in this field already have good eyesight. When the scallops are difficult to spot it is best to swim slowly and close to the bottom, stopping periodically to reassess the ground. When they are easily spotted, however, a distance of one to two feet from the bottom will enable a greater expanse of ground to be seen.

It must be remembered that a good deal of mental planning goes into scallop diving, and as with any skill, it is difficult for anyone to analyse just how they acquired it. A simple problem will illustrate this: on sparsely populated ground, five scallops are spotted at once, covering an area of about thirty square yards. In which order should you pick them up, to make sure that they are all collected? The answer is that they must be systematically picked up, in such a way that the ones

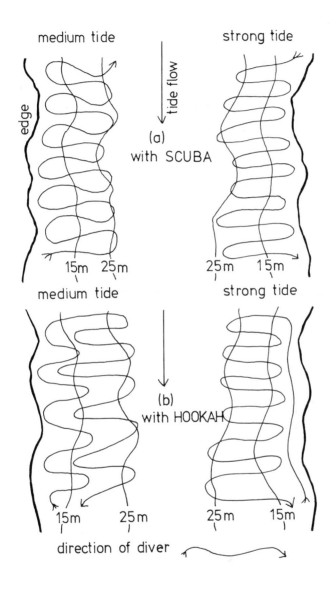

Fig 10 Search patterns when looking for and gathering scallops, showing how tide and air supply affect the method of working.

remaining are not obscured by sediment. Whether they are aware of it or not, good scallop divers put much thought into what they are doing, and subconsciously plan their approach to fishing.

Once one scallop is spotted, another must be looked for, and so on. With practice a diver can pick up one scallop after another without ever ceasing his search. He must be looking all the time, even while in the process of picking up a scallop. One good tip given to me when I first started scallop diving was, once a scallop is spotted, look for its mate. The point of the advice was to keep concentrating and keep your eyes open all the time. One very obvious, but nevertheless good, point to improve fishing is to put two or three scallops in the catching bag at once. Movements are very much impeded on the bottom, so the fewer made, the less effort exerted, and the more efficient the operation. Personally, I pick one scallop up and use this to scoop the second and third, before putting all three into the catching bag (*Fig 11*). If this simple exercise were to increase a diver's catch by as little as a dozen scallops a day, then it would be worth it. It is sometimes an advantage to count the scallops into the bag, as this gives a scale to work to and can result in fractionally better

Fig 11 A diver working at 60ft picking up scallops.

36

catches: if the catch is not up to scratch after the first half of the dive, then you work all the harder to achieve your target by the end of the dive.

Scallop divers will rarely agree over methods of working on the bottom, but the points mentioned so far should give some guidance to approaching the job. Many divers I know would never think of working sparse ground and are always looking for the thick patches. This is fair enough but it tends to limit the skill of the diver as a fisherman. To become a good fisherman he has to learn to have patience, and working thin ground is good training for this.

Summary

To summarize this chapter, a good scallop diver should have the following attributes:

i) he must know where the fishing ground is, or where it is likely to be, and must remember areas that produced worthwhile catches for future reference.
ii) once he has reached the bottom, he must know where to look for the greatest density of scallops, including on ground that has recently been fished by someone else.
iii) he must have keen eyesight and be able to spot scallops at a distance.
iv) he must be a fast and efficient under-water worker.

Finding and gathering scallops is a skill that is acquired over a period of time, and no one can hope to master it immediately. Divers just coming into scallop fishing will benefit from the knowledge of established divers. When diving for scallops first started, divers ignored certain areas completely because they thought they knew the exact position of all the beds. This proved a costly mistake, and many divers actually gave up because they thought all the areas had been fished out. By looking further afield and giving more thought to the job, it was realised that the scallops were far from finished. Consequently, many of the former ideas about scallop diving were abandoned. Divers now look on the job as a profession instead of as a means of making occasional quick money. Anyone who perseveres and masters scallop diving will always have work, but the process of learning can often be hard and disappointing.

4 The diver's boat

Once a decision has been made to fish for scallops on a full-time basis, thought must be given to the type of boat to be used. There is no need for everyone to have his own boat, as some divers are content to work on a share basis, but it is still worth knowing something about boats in case the need to buy one ever arises.

Types of boat

There are many different styles of boat on the market but the fishing industry has, by trial and error, narrowed its choice down to a few basic types. Plenty of beam and a reasonable draft are essential for a fishing boat, be it open or decked. The older, traditional, boats tended to have a wheelhouse sited aft but many of the newer boats have forward wheelhouses. Hulls may be of wood, GRP, steel, ferro-cement or alloy. Whatever the material, the hull must be strongly built, and if made of wood must be substantially framed. With GRP hulls, always select the heavy duty models, which are substantially stronger.

Scallop diving can be carried on from almost any type of boat that floats (*Fig 12*), but over the years attention seems to have been centred on three distinct types. First, there has been the small, fast lightweight craft of the inflatable and sea sledge type, which has the advantages of being speedier on journeys to and from the scallop grounds, having relatively low setting-up costs, and providing a certain amount of independence (*Fig 13*). Second, there is the traditional style fishing boat of 25 to 35 feet, with either forward or aft wheelhouse which gives more scope to the diver and is an agreeable size of boat to operate (*Fig 14*). However, such a boat does involve a greater commitment and also a higher capital outlay. From here there is a natural progression to the third class of diving boat, the craft of about 40 feet and over. This category involves an even greater capital outlay with very much increased running costs, due to the strict regulations imposed by the Board of Trade on boats over 12 metres (39.38 feet).

The choice of boat will depend upon the amount of money available and the commitment desired. Obviously a boat of 40 feet and over will involve a considerable commitment, yet the returns can be very good if the right divers are employed. It seems that the trend in scallop diving boats will follow that in the abalone industry, where the type of boat now preferred is 18 to 23 feet long,

Fig 12 Four scallop diving boats, showing the variety of craft that can be successfully used in the same area.

Fig 13 The 'Dory 17', a fast lightweight petrol-engined diving boat.

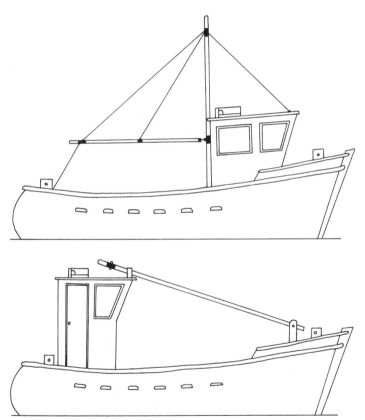

Fig 14 Typical small fishing craft: (*a*) with wheelhouse forward; (*b*) with wheelhouse aft.

41

with two 110hp outboard engines. This has enabled divers to fish as far as 40 miles off the coast, and to travel there and back in one day. Such a boat, because it gets to and from the grounds faster, gives more time to fish and opens up fishing grounds beyond the range of less powerful craft. A further advantage is that the boat is easy to transport. A journey to a fishing ground that might take many hours and many gallons of petrol by sea can sometimes be made in half the time and at half the cost by road. Similarly, the boat may be transported to a sheltered site if the diver is weather-bound on his normal ground.

Unfortunately, these small high-powered boats are expensive to run. Their engines are costly and, because they are highly-tuned, not very reliable. Both the boat and its engine depreciate quickly.

To own or to share?

In the scallop diving business the most common arrangements with boats are the following: to work on a share basis, hire a boat, enter a partnership with one, or own one outright. There are various advantages and disadvantages in all these arrangements, and one point deserves consideration above all the rest – the responsibility of owning or running a boat.

Boats are very time-consuming to operate and maintain and one can easily become tied down by their need for constant attention.

Points to look for

Buying a boat is a big step financially, and this can often lead to worries, especially during periods of poor returns. However, as long as there is an awareness of the restrictions imposed on one by the ownership of a boat, then one should go ahead, bearing in mind the following points: the boat's seaworthiness; its manoeuvrability; its adaptability for other types of work; its chances of resale; its ease of maintenance; its running costs; the earning potential as a ratio to its net capital outlay.

Seaworthiness and seakindliness

If a boat is not seaworthy many diving days will be lost. Being 'unseaworthy' does not necessarily mean that the boat is rotten or inadequately maintained, it can also refer to its unpredictable and dangerous movements in rough sea. If it rolls a lot in bad weather, or perhaps ships a lot of water, then it is of little use as a diving boat. A boat of narrow beam for its length, and of shallow draft will not be a good sea boat. Diving necessitates much loose gear being on deck, so a boat that remains reasonably stable in rough weather is essential. To ensure seaworthiness and avoid poorly maintained or dilapidated boats, a qualified surveyor should be engaged to advise on whether a boat is worth buying or not. In any case, some kind of survey will be needed to satisfy an insurance company that the boat is seaworthy.

Manoeuvrability and visibility

Manoeuvrability is important in a diving boat because it is not always easy to come alongside a diver in the water, especially in rough weather. Furthermore, a diving boat can be quite manoeuvrable and yet be handicapped by poor visibility from the wheelhouse. It is important to have a combination of good manoeuvrability and good visibility (from the wheelhouse), in order to avoid mishap when coming alongside a man in the water. A boat can be made more manoeuvrable by altering the shape and size of the rudder, and even greater manoeuvrability can be attained by fitting a nozzle around the propeller. These refinements will improve the boat's handling properties, but an important point to note is how the boat behaves when the propeller has stopped turning. Some craft will carry on under their own 'way' and respond to changes in course quite readily, while others will not respond at all to the rudder until the propeller is turning again. This is a very important point because the propeller should have stopped turning well before the boat has reached the man in the water. As to visibility, it is perhaps better to buy a boat with a forward wheelhouse rather than aft, because this gives the man at the wheel a far better chance of seeing all that is going on.

Adaptability

It is pointless buying a boat if it cannot be adapted to some other type of work. When scallop diving, there is always a chance that an offer or opportunity will occur in some other field, which, if followed up, could prove profitable, so the boat should be adaptable to changes in work. A mechanical or hydraulic hauler is a big advantage on any boat because this will enable it to turn to many forms of inshore fishing (*Fig 15*). Plenty of clear deck space is essential, and lifting gear, such as a derrick or boom, is a great advantage (*Fig 16*). Although alternative forms of fishing may not be desired as an occupation, it is an advantage to be able to turn to them in times of necessity.

Resale value

In addition to its versatility, a boat that can adapt to other types of work is a great advantage if ever there is a need to sell it. There is a limited market for diving boats as such, but small fishing boats are popular. The boat should never be allowed to deteriorate to such an extent that its resale value will be affected. A boat is a large item of capital outlay, and if looked after properly should appreciate in value.

Maintenance

Maintenance is something that every boat requires constantly, but a wise initial choice can save much time in the long run. Some of the newer types of hulls made of GRP or ferro-cement are economic in general maintenance and repair. Steel and alloy hulls are also time-saving, but the former should be regularly scrutinized for excessive corrosion. The choice of engine also greatly affects the amount of time spent on maintenance. Some older engines

Fig 15 A forward-wheelhouse fishing boat used for scallop diving. Note the hydraulic hauler beside the wheelhouse.

Fig 16 An aluminium assault craft. Strong, light and inexpensive, it can be powered by any size of outboard and makes an excellent diving platform. Note the derrick to help in raising the catch.

require much more diligence and patience than the more modern higher-revving engines. Some engines are very heavy and some are relatively light, and this difference is soon noticed when the power unit has to be taken out for repair. In addition, the slow-revving, heavy engines are much harder on the hull, causing greater vibration than the lighter, higher-revving models. Maintenance should never be skimped, but steps should be taken to ensure that it is kept down to a reasonable amount. After all, the effort of having to maintain the diving gear efficiently is enough in itself. Careful planning and a certain amount of preventive maintenance should ensure that any major work is kept to a minimum.

Running costs

The running costs of a diving boat need close scrutiny. There is nothing worse than having to meet a high weekly outlay even when the boat is tied up, so anything that can be done to cut down costs must be considered. Above all, prospective returns must be estimated before buying. If they are likely to be poor then it would be better to put the initial capital into something giving a greater return. It is important to study carefully the ratio between gross capital outlay and net returns on an annual basis.

5 Methods of working

Factors affecting methods of working

Scallop divers will argue endlessly among themselves about methods of working, and each diver usually becomes more and more convinced that he is right as the argument goes on! The trouble is that it is easy to generalize about good working methods – they should be safe, efficient, profitable and simple – but impossible to say what is best for one diver or team without knowing just what their circumstances and personal preferences are.

Apart from safety, which must be the main consideration, there are four important factors to take into account in deciding how to organize the work of diving; air supply, size of boat, number of divers, and how the catch is to be lifted and landed. It will be seen that none of these factors can be considered in isolation; each of them has a bearing on the others.

Air supply

The diver's air supply can be self-contained (scuba) or via an air line to a low-pressure compressor on the surface. The low-pressure system is cheaper, takes up less space and is more portable than a high-pressure compressor and air bottles, though scallop divers have been slow to realize this. In contrast, the abalone divers of Australia and

California dispensed with high pressure air as soon as there was a trend to one- and two-man diving from small craft, where the advantages of low-pressure air are greatest (*Fig 17*).

It was not uncommon for only one member of a two-man scuba team to be able to work at a time when diving from a small craft. There is little room to spare in such a boat and it was sometimes not possible to carry more than one set of gear. Also, a scuba diver can have difficulty in getting aboard a small boat unaided. A low-pressure compressor takes up little room in the diving boat, the divers

Fig 17 A typical Australian set-up for two-man abalone diving with low-pressure air supply. Catching bag and 'floater' are being brought aboard.

are less burdened with equipment, and both can work on the bottom at the same time.

The disadvantages of low-pressure air are that the boat must be anchored, and that the diver's range is restricted to the length of his air line. Neither of these considerations is very important in one- or two-man working, but they impose serious limitations if working with a larger team, when it would be necessary to haul the anchor and move the boat many times in a day to find enough ground for the divers to work. For this reason, larger teams of divers prefer scuba gear.

The air line of a low-pressure system can be a positive safety factor, providing a permanent link with the surface, which does not exist with some other forms of diving. Unfortunately the line is also vulnerable to damage, notably from propellers of passing boats.

Type of boat

It is now quite usual to dive for scallops from small craft, but vessels up to 40 feet and more are in use, and every boat presents the diver-fisherman with opportunities and limitations related to its size.

Small boats cannot provide all the protection a diver would like for all-year-round fishing, and larger boats can put to sea in worse conditions (though only small boats can be transported to sheltered water).

A boat about 30 feet long will accommodate three divers and their scuba gear in comfort and is a usual next step from a small boat. But remember – such a change of boat will require a complete reorganization of the diver's work. The usual method of working for a three-man team is for one member to tend the boat while the other two dive.

The transition to boats above 30 feet in length is relatively smooth, but as boats get larger they also become less manoeuvrable, and unless a boat can approach a diver in the water safely it may be possible to use it only for travelling to and from the fishing grounds, diving having to take place from inflatables – a very expensive arrangement.

Number of divers

The close relationship between the size of a diving team and the size of its boat is perhaps clear enough. What may not be quite so obvious is that having to organize a number of divers and a large boat involves administrative and financial worries, quite unknown to one- or two-man divers, that can completely alter the nature of the job.

It may seem that a diver working alone from a small craft must be less safe than if he was a member of a team. In fact a lone diver tends to build more safeguards into his working methods than do many larger units.

While a team of divers can cover more ground in a day's work than one diver, it is far easier for one man to work a ground thoroughly than it is for divers in a larger team, put off at random over a greater area.

Lifting and landing the catch

Since scallop diving started there have been many changes in diving techniques, aimed at getting the diver the best return for the least time and effort spent underwater. However, old methods die hard, and inefficient and costly methods of getting scallops from the sea bed to the boat and from the boat to land are not unknown today.

The most exhausting method of lifting scallops would be to swim to the surface with full catching bags, a method employed by early abalone divers. An enormous improvement is simply achieved if a buoyed line is attached to the catching bag, which can then be left on the bottom when full, the diver surfacing along the line, descending again with an empty bag and new buoyed line to the spot he was working, the full bag then being lifted from the boat.

This bag-and-line method has much to recommend it: beside saving effort, the buoyed line marks the diver's position which, if he was scuba equipped, would otherwise be unknown. It enables him to return easily to the ground he is working, and his frequent trips to the surface keep him in touch with the boat.

A further saving in effort can be made by using a buoyancy aid to lift the catch. Abalone divers used a PVC 'parachute' but scallop divers have adopted a rigid 'floater', usually a 5 gallon plastic drum. The buoyancy aid is attached to the catching bag and is full of water during the descent but topped up with air from time to time as fishing proceeds, to support the increasing weight of the catch (*Figs 18, 19*). It enables heavier catches to be

Fig 18 A scallop diver working at 60 feet, with catching bag and 'floater'.

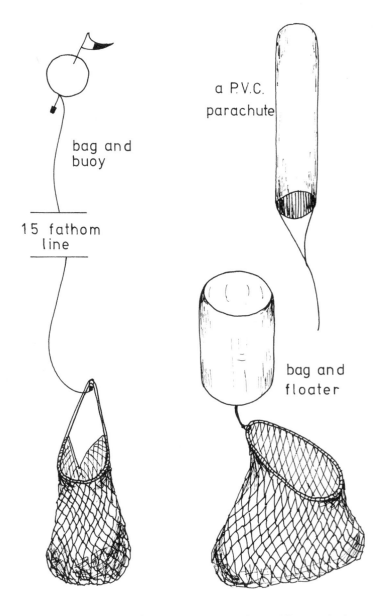

a P.V.C.
parachute

bag and
buoy

15 fathom
line

bag and
floater

Fig 19 Three methods of raising the catch: (*a*) buoyed line attached to
catching bag; (*b*) a PVC 'parachute' buoyancy aid; (*c*) a rigid plastic
'floater'.

lifted and cuts down the number of trips a diver must make to the surface while he is fishing.

It is possible to combine the greater safety of the bag-and-line method with the efficiency of the floater system if the diver using a floater also carries down with him a buoyed line to mark his position in the water. When he surfaces the line can be clipped to whatever weighty object is available on the bottom, to guide him back to his fishing ground. There is a danger that when a diver surfaces at the end of a dive and the line lays slack on the surface, it may become entangled with the propeller of the diving boat. This can be avoided if the diver attaches the end of the line to a small weight before he comes up. The line is thus kept relatively taut, and can be hauled into the boat after the diver is aboard.

Keeping the catch

Most diving boats try to market their catch every two or three days, and so avoid the trouble of putting the scallops back in the water to keep fresh. There are certain circumstances, however, that will justify them doing this, such as having to stay out for four or five days at a time, or when bad weather prevents a landing. In such cases it is best to be prepared to 'shoot the catch away', and there are certain rules that should be observed when doing this. If the scallops have been out of the water for too long, they will die as soon as they are put back in, because of the effect of the sudden temperature change on their already weakened state. Therefore, the diver must try to get his catch back in the water as soon as possible.

There is a danger of contamination if the scallops are 'shot away' in an area that is exposed at low water. This is particularly true in areas where there is a lot of shipping, and therefore oil lying on the water. Generally speaking, the scallops should be put in an area of strong tide, which is deep enough to keep them covered at all times. Small mesh net bags are favoured for keeping them in, and it is better not to pack them too tightly, allowing them to move around in the bag. If the mesh is too large predators such as star fish, dog whelks and crabs will enter the bag and severely deplete the catch. On the other hand, if the bags do not let an adequate flow of water through, the scallops will soon die.

Whatever method of working is chosen, there will always come a time when scallops have to be stored back in the water. With a mechanical hauling aid fitted to the boat it is easy to lift the bags on to the boat, and in this case a good number can be spaced out on one main line so that all can be hauled at once. However, not all boats are so equipped, and in most cases the full bags will have to be put back into the water on single ends, each end being tied to a main buoy (*Fig 20*).

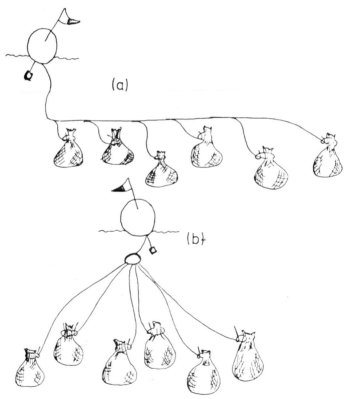

Fig 20 Keeping the catch: keep-bags 'shot away'; (*a*) in a fleet; (*b*) as single ends.

Summary

Scallops divers will always argue among themselves that each has the best method of working! The arguments usually favour one air supply versus another, a fast boat versus a slow boat, or one type of buoyancy aid versus another. The arguments cannot be conclusive: they only serve to make each diver more convinced that he is right. One way of working may suit one man and not suit another. Some divers do not want the financial complication of running a large boat and employing a number of divers; they would sooner work on their own, with fewer worries. Other divers do not want to get involved at all in running the job, and are content to be employed by a diving boat on a share basis. The only real criterion of any method of working is for it to be as safe and efficient and profitable as possible, without it becoming so complex that it is no longer enjoyable.

6 A working set-up

In this chapter details of methods and equipment for a complete working set-up, from the boat to the marketing of the catch are given. It must be remembered that it reflects one man's opinions; there are many possible variations and the choice between them ultimately depends on the personal preferences of the diver.

The boat

Boats of about 30 feet in length, with their wheelhouses forward, have been popular with scallop divers, so I will base my working example on this style (*Fig 21*). One very important feature of this kind of boat is the self-draining deck aft of the wheelhouse. This increases the craft's potential, especially its ability to operate in heavy seas. Particular attention should also be paid to the fo'c'sle, because if men have to stay together for any length of time they should be housed as comfortably as possible. Three bunks should be adequate and there should be plenty of locker space. A good propane gas stove and a gas, solid fuel, or diesel heater are essential.

The working capabilities of the boat will be determined by seaworthiness, the amount of clear working space available, and any mechanical hauling aids that it might have. Seaworthiness will be determined by the craft's breadth and draft; be sure that it has plenty of both. When necessary, extra ballast will help to keep the boat on an even keel in rough weather. The availability of working space is very important as there is nothing worse than trying to work on a cluttered deck. All available space should be fully utilized and any unwanted items stowed in the hold. Mechanical hauling aids are a big advantage on any boat because not only do they reduce the amount of physical effort needed to haul things off the bottom, but they render the boat useful as a fishing boat as well as a diving boat. This is important when it comes to selling the

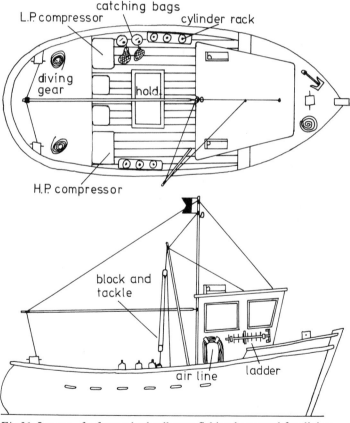

Fig 21 Layout of a forward-wheelhouse fishing boat used for diving.

boat or when it is decided to take a break from diving. Fishing may not be the first choice of occupation, but it is an alternative, if necessary.

The question of running expenses is one of the most important points to consider. A few pounds saved each week can represent a great deal at the end of the year, and may make the difference between a profit or a loss. Not so many years ago the amount of diesel fuel used by an engine was not important, but nowadays it can be critical. Big, greedy engines are too expensive to run, especially if the extra power is not really needed. Most displacement hulls have a maximum speed, and it does not matter how big the engine is, above a certain size, because the boat will not go any faster. A reliable diesel engine of forty to fifty horse-power is the ideal size for a 30 feet long boat, and if this power is used properly through a reduction box, the boat will have power to spare. Other items of expense are insurance and the hire of electrical gear in the wheelhouse. Nobody would suggest cutting down on insurance, but it is worth shopping around for the cheapest quotation. Some fishermen like to hire such things as echo sounders and radios. This can be burdensome, and if added to insurance can amount to a fairly heavy weekly outlay, the more so when the boat is laid up for repair. With electrical equipment, I consider it best to buy outright or try to get some second-hand. By keeping expenses low there will be less to worry about when the boat is not earning.

Diving gear

The next point to consider is what kind of, and how much, diving gear to carry. I have opted for a three-man operation, with provision for four, if desired, and even with a small unit like this it is amazing just how much gear is needed. If the divers on board are roughly the same size, then many items can be interchanged, but it is still advisable to have more than you need. There is nothing worse than being held up for the sake of a lost mask or something equally trivial.

The first major item needed is the compressor. This

would be a high-pressure type, because I would opt for scuba gear as a means of working. A high-pressure compressor is a very expensive item to duplicate, so, as a compromise, it is well to have a low-pressure system available to ensure a reliable alternative supply of air. The low-pressure system is also very useful if decompression is required, by going back into the water. The high-pressure compressor should be of a reasonably high output (5 to 7 cubic feet per min) at a working pressure of 3 000 psi (pounds per square inch), and it should be as light and as portable as possible to enable it to be taken off the boat easily if repairs are necessary. When stationed on the boat it should be well protected from the outside elements, and the air intake should be positioned well forward of the exhaust. If possible, the compressor should be driven off the boat's main power supply as then, not only is time saved in filling the air bottles but it is done more economically. Unfortunately, this is not always practicable because of the lack of space in most engine rooms. By far the best independent power unit for driving a compressor is a small diesel engine. Although more expensive than a petrol unit, it will last much longer and should give far less trouble. To save some of the time lost through compressor breakdowns, it is good policy to fill the cylinders immediately after they have been emptied, in preparation for the next day's dive. This ensures that if any trouble is met, there will be some extra time to put it right before the next day.

The next item on the gear list is air bottles. It is usual for each diver to have about two hundred cubic feet of air at his disposal, broken down into a twin pack and a single. For three divers, therefore, nine bottles are needed, of about seventy cubic feet each. Twin packs are labour-saving when it comes to filling, and it is also an advantage to be able to work with such a large amount of air in one dive. If, however, single bottles are preferred, then a quick-release harness for use on all of them will help make the whole operation much easier. One spare bottle above the quota should be carried just in case of emergencies, and care must be taken to make sure it is always topped up. If possible, all the bottles should be of the same size and brand, then manifolds and harnesses are interchangeable, and spare parts are easier to order.

When working with three divers it is best to carry four demand valves, each with its own contents gauge and low-pressure suit inflation hose. This may sound an unnecessary expense but if ever one stopped functioning and decompression was needed, then the spare would be essential. As to the rest of the diving gear, such as fins, masks, weight belts, knives, gauges, *etc*, three sets of each should be sufficient because most of these items are interchangeable.

The question of how many diving suits to carry is difficult. Most scallop divers prefer a dry suit worn over a good undersuit, but this is expensive to duplicate. It is therefore better policy to carry a wet suit as a spare, because it is an all too common experience to get dressed up in a dry suit, and then find that one of the neck bands is lost or the cuffs are too badly perished for it to be usable. A spare undersuit is a good idea, especially if the dry suit is prone to leaking. There is nothing more uncomfortable than standing around in a dry suit half full of water, and this could lead to a diver calling off the rest of his work for the day. With a spare undersuit he can be dry and comfortable again in a matter of minutes, and not be too daunted by the prospect of his next dive. For three divers therefore, it is good policy to carry three dry suits and a spare wet suit. One important point to remember here is never to wear a dry suit on a boat without the neck seal done up. A couple of extra neck bands and plenty of good spare neck seals are essential. The spare wet suit can be an inexpensive one, as it will only be used if and when the dry suit needs repairing. If the dry suits are well looked after and periodically inspected for perishing and excess wear, many delays can be avoided. In case it becomes necessary to use the wet suit, a life jacket should be carried, which must be checked regularly and stowed where it is unlikely to sustain any damage.

The last items of diving gear are catching bags and floaters, or buoyancy aids. It is good to have a few of these to spare because one may be wanted in a hurry, and also it is a nuisance to be held up while someone mends a hole in one. Catching bags are easily made, and buoyancy aids can be bought quite cheaply from hardware stores.

When all this gear is loaded on to the boat, along with other items that the boat must carry, there will not be much room to move unless it is stowed away securely to leave as much working space on the boat as possible. Racks should be made up for the bottles, and lockers for the rest of the gear, with everything being secured in such a way as not to break loose in rough weather. It is useful to have somewhere to hang up suits to dry, such as a line run off the mast. A diesel burner in the forecastle will give off enough heat to dry the suits quickly, and, if desired, it can be left on all night quite safely. With the heater turned down low the amount of diesel used is minimal.

Methods of fishing

The next point in the working set-up is the method of fishing. This is very important and a logical approach will pay dividends in the long run. The most common method is to use the bag and floater, and a safe refinement to this is to have a line and buoy attached to the catching bag. The line should be such that it can be quickly unclipped from the bag and fastened on the bottom when the diver's time is up or his bag is full.

Building up a knowledge of the ground requires years of effort but short cuts can be taken, by listening to other divers, seeing where they are working, and recording how successful they are. This does not mean that you should move into their area immediately, but if you make a note of their progress you can learn whether the particular piece of ground has a covering of scallops or not, and so, whether it is worth remembering for a future date when the ground has been repopulated. A record should be kept of where and when all catches were made, no matter how small, and by a simple form of elimination a general outline of an area will emerge, and the boat should spend less time on infertile ground.

61

I find it a good policy to work two or three pieces of ground at a time, moving from one to another regularly. In this way, the grounds tend to get worked more thoroughly, and it is less tedious than being stuck on one particular patch for weeks on end. No matter how well a diver thinks he has fished a piece of ground, someone will always come along and find a piece on it that was missed. This is why it is a good policy to move in rotation from one ground to another thus approaching each piece with a fresh view irrespective of its having already been worked or not.

When starting to work off-shore ground, it is a good idea to mark it out before progressing further. Once it has been located on the sounder, run over it a few times to get the layout, then mark the ground with buoys at its extremities. Three markers are usually sufficient and, if possible, use dan buoys with 'diver down' flags attached to them. The diving can then be carried on inside the markers and, if the particular piece of ground is to be worked for a number of days, all the markers (but not, of course, the 'diver down' flags) can be left out, including the diver's personal line marking the position of his last dive. In this way much time is saved, and it also means that the ground is worked more thoroughly. Any piece of ground that is found a long distance off shore should be marked on the chart, along with fixes taken off the surrounding land.

Financial arrangements

The setting-up of the boat and gear can either be done as a partnership or as a one-man enterprise, with other divers working on a share basis. On a share system it is usual for the employed diver to forfeit 25% of his catch to the boat, and for the expenses to be split evenly amongst the number of men aboard the boat. The employed divers will supply their own gear, but their bottles will be filled by the boat's compressor. This is a better way of working than a partnership, because it gives all the men on board some form of independence. It also makes it possible to get some new ideas introduced to the job. A new man to the job, no matter how inexperienced, will often put forward a suggestion that will improve the fishing in one form or another. It is always the simplest things that are overlooked, and a diver just starting the job will spot them more quickly than the man who has been doing it for years.

Landing and marketing

The final point in our working set-up is the question of marketing the catch. Ideas on this have changed over the years, as scallop divers have become wise to the workings of processers and fish merchants. There was a time when a boat would stay with one particular outlet almost indefinitely, but this tended to limit the price range, so now the policy is to distribute the catch between two or even three buyers, thus securing a wider market. Healthy competition among buyers has not only helped to push the price up, but it has also helped to keep it more stable.

Most scallops are sold at a price per shell. Four and a half to five and a half inches is a medium size, and most scallops fall within this category. Prices rise dramatically, however, in the large shell group of up to six and a half inches and over, and this has tempted many divers to concentrate their work in very shallow water where the larger scallops are usually found. From three and a half to four and a half inches is considered a small scallop, and, unfortunately, processors offer a price for each shell in this range. Most divers agree that it would be advantageous to the industry if this size of scallop was left on the bottom to grow and to reproduce for the future.

The alternative is to sell by the weight of meat per shell, but then there is no way of the diver knowing what weight he has landed until the scallops are actually processed. If it is more convenient to land by meat weight alone, it is very important to leave the scallops in the water until the very last moment. Two to three days out of the water can sometimes mean a two to three pence per shell loss in meat weight, caused by dehydration.

With only three men working from one boat it is worth

storing a few days' catch at a time, in order to make the landing more worthwhile. Because of transport costs fish processors do not like sending for a small quantity of shells. Larger landings provide a far better bargaining position. Two landings a week is usually sufficient, and three divers should be able to manage about 2 500 scallops between them on each landing. This represents a reasonable size of catch, and, if landed in conjunction with another scallop boat, will provide a more attractive load for the processor or fish merchant.

Accounts

One essential point in the working set-up is the necessity of keeping neat and accurate accounts. All transactions should be methodically recorded, and it is good policy to keep a book on board the boat at all times to note incidental expenses and landings. All receipts should be filed, and a petty cash book be kept accurately and up to date. By keeping accurate books the accountant's job is made much easier, and consequently less costly.

Summary

A decision must be made as to the extent of commitment to scallop diving desired, because a boat such as the one described will certainly require heavy physical and financial involvement. The most important thing, apart from receiving a steady income from the investment, is to make sure that the initial capital outlay does not depreciate too rapidly. Obviously, diving gear does not hold its price for long, but a good boat, well maintained and fitted out, should always represent a reasonable amount of capital.

This set-up can, of course, be either scaled up or down depending on requirements, and either way it is important to put thought and effort into the system and run it in as efficient a way as possible. After a while this kind of approach to the job will show good returns and it will also give more working satisfaction.

7 Diving alone

Both the Australian and Californian abalone diving industries owed their spectacular growth to one-man diving operations. The same has not been true of scallop diving although a small proportion of divers have opted to work completely on their own, possibly influenced by those Australians and Americans who have spent time diving in this country. These lone divers were efficient and very safety conscious, so far as was possible when working in that manner.

Legislation at both club and government level in the British Isles now prohibits any type of lone diving, especially for profit. There are good reasons for this but, on the other hand, one-man operations have had a very safe working record, most of the fatalities being the result of shark attacks in such areas as Australia and California. Obviously, there are hazards in working under water totally alone. However, as diving alone has played an important role in the development of all types of shellfish diving throughout the world a brief outline of the practice is given below.

Preparing to dive alone for scallops was more interesting to me than organizing a two- or three-man diving operation because many new problems were encountered and resolved over a period of time by trial and error, and the end result was a fast and efficient system of working which was both labour-saving and economical.

When there was no one to help with all the carrying and transporting of equipment, it was obviously necessary to keep it to a minimum.

Safety

The overriding concern with a one-man set-up was to work in as safe a manner as possible, yet to keep all safety aspects at a manageable level. The worst problem the diver faced was the activity of boats overhead while he was on the bottom, and this risk was aggravated by the natural curiosity people have about an apparently empty boat. A 'diver down' flag was not really enough indication of what was going on, so, where possible, the word 'DIVING' was painted on each side of the boat in bold letters. Fishermen soon became aware of the activities of one-man scallop diving boats and gave them a wide berth, but the same could not be said of private boat owners. No one could be blamed for being curious, so the safest thing was to make sure that the visitor was made aware of what was going on from a good distance away.

Air supply

As most one-man diving operations used a Hookah air system (surface supplied air), there arose the problem of the air hose floating on the surface of the water where it could easily be picked up by a boat's propeller. To combat this danger air lines were, and still are, made in a colour that is easily spotted in the water. However, this remained a problem when there was a lot of boat traffic about. Another way of reducing this risk was for the diver to use a line and buoy, with a 'diver down' flag attached to the buoy. By working this way, any approaching boats could easily make out the position of the diver in relation to his diving boat, and consequently avoid passing between them.

The diver's air supply presented another problem. If the compressor broke down whilst he was in the water he would have to surface immediately, using the reservoir of air stored in his hose. This was quite a straightforward manoeuvre in normal conditions, but could prove dangerous if the ascent had to be controlled for purposes of decompression. A simple way of avoiding this problem was to introduce a high-pressure air bottle somewhere into the system, being careful to ensure that the air pressure was reduced to that in the low-pressure compressor. By introducing this bottle at the compressor end a special valve was essential which would open when the air in the machine fell below a specified level, thus replenishing the system with air from the bottle. Possibly the safest way of utilizing a high-pressure reserve of this nature was to actually wear the bottle whilst in the water. The air line was fed into the low-pressure side of the demand valve via a quick-release snap connector, and if the bottle were worn in an inverted position, the reserve could easily be turned on if there was a failure from the surface. A non-return valve at the diver's end of the air line was an important addition to this system. In cases where the line burst, the valve stopped the air from rushing back up the line and straight into the atmosphere.

Planning the work

Some divers operating on their own made extra work for themselves by not putting enough thought into planning their activities. When I started on my own I used scuba bottles and filled them at home, usually providing four days air at a time. Apart from having to manhandle these cylinders backwards and forwards from boat to home, I also had to put petrol aboard every couple of days, and to land the catch. Most of my effort was being consumed in preparing my boat for sea, and by the time I got into the water I was in no fit state to work efficiently. It soon became obvious that a new and improved method of working needed to be developed.

By introducing a surface air supply the problem of transporting and re-charging the air bottles was overcome, and this proved a big step in the right direction. The next problem was to try to reduce the amount of handling involved in landing the catch. Carrying sacks of scallops up a beach each day was not conducive to good temper. The answer was to try to keep the landings down to one a week. This one large landing could be put ashore either when the tide was right, or when adequate facilities were organized for its despatch. In order to keep the scallops fresh they had to be put back in the water soon after they were caught. By 'shooting the catch away' in this manner, the daily work load was eased considerably, and it then became pleasantly apparent that the main task was to dive. This was the stage that all one-man set-ups tried to reach; where the minimum amount of physical effort was needed to organize the boat for sea and to market the catch. At this stage the diver could turn his attention to organizing his diving so that, first of all, his returns were satisfactory, and second, that he worked in a labour-saving way. Working alone out of a boat was very tiring, but it was surprising what results each operation could produce with careful forethought.

The boat

The size of boat was important to the success of the operation. It had to be light enough to transport easily, yet large enough to be a stable platform to work from. It was not uncommon for divers to work on their own out of much larger boats than mine, but in these cases the running costs were usually too high to enable the full rewards of the job to be obtained.

A one-man diving set-up

To give an idea of how these one-man outfits operated I will outline a typical set-up, comprising a 16 feet long fibre-glass boat with a strong trailer, a forty horse-power outboard motor, and a low-pressure compressor with 300 feet of air line.

When equipping a small boat like this for sea, care had to be taken not to clutter the working space. The basic diving gear of fins, mask, weight belt and demand valve

68

did not take up too much room, and incidentals such as decompression meter, knife, depth gauge and hood could be stowed away easily. Three hundred feet of air line is a bulky item, but was stowed in such a way as to permit ease of handling by always flaking it down into a tight figure-of-eight shape. The catching bag and floater, net keep bags, extra rope and marker buoys were stowed away tidily in one large net bag. In addition to all this room had to be found for the anchor and cable, five to ten gallons of petrol, two-stroke oil, a tool kit, a pair of oars, and a couple of dozen sacks. In some cases it was convenient to carry a second outboard motor (usually a small Seagull), mounted on the stern of the boat as a standby. The final item was a 'diver down' flag, flown in a high and prominent place.

It was usual practice when working alone to get kitted up before going aboard the boat. Most one-man operators opted for dry suits or uni-suits, but, unfortunately, the assistance needed to get in and out of the latter was not always available. On the other hand, with practice it soon became quite easy to get dressed in a dry suit with no help. Having checked that all the gear was aboard, and being suitably kitted out for the water, the diver was ready to look for scallops.

Spot diving for new ground was more complex with this kind of one-man set-up than it was with a conventional diving boat, so careful thought was put into where the search should start. By working ground systematically, taking the good with the bad, much spot diving was eliminated, but a good understanding of the ground was a prerequisite of this procedure. Echo sounders were sometimes carried on these one-man boats, but because of the likelihood of damage more emphasis was put on the use of charts. Once a piece of ground was chosen, the boat was anchored and the outboard tilted out of the water (to avoid snagging the diver's line). The compressor was started and the air system checked over. Most connectors were of the quick-release type, to speed up the process of connecting and dismantling gear. The demand valve was connected to the air line by a one-way snap connector and the suit inflation whip was connected in the same way. Stainless steel clips were used to attach the air line to the diver's harness, and also to the catching bag once it was filled with scallops. This enabled it to be hauled in by the line.

Divers found the best way of working with an air line in areas of strong tide was to try to work into the tide for the first stage of the dive, zig-zagging ahead all the time, and then falling back with it during the latter stages (*Fig 10*). It was a good idea to try to surface close to the boat but this didn't always work out in practice, and often the diver found himself a full hose length away. Once the bag was full it was floated to the surface, followed by the diver, and then clipped on to the air line. The diver freed himself of line and valve and proceeded to swim back to his boat by way of the air line, which was now lying stretched out on the surface. Once the diver was aboard the boat the compressor was stopped and the bag hauled in. In areas of extra strong tide it was easier to cast off from the buoyed anchor and fall back on the catching bag, thus eliminating a particularly strenuous exercise. Once the bag was hauled in, the scallops were counted into their keep bags and shot away on lines in a suitable place.

Advantages and disadvantages

The main advantages of working alone were: the total independence it offered, the shorter working day, and the relatively low cost of setting-up the whole operation. Having the independence to operate in exactly the way one chose was something worth having, but it required a degree of self-discipline to go out to work on a cold winter morning. However, this independence also meant that the boat could be laid up at any time without one having to worry about anyone else's livelihood; also the fact that it could be stored in a convenient place on dry land eliminated any worry about its safety in bad weather. The prospect of a relatively short working day was to many divers a great advantage. The whole operation could be completed in under five hours if all went well, but this, of course, was dependent upon a fair amount of time and motion study by the diver. Over a period of time the pitfalls of this type of operation could be resolved, but they first had to be experienced to enable a sound alternative to be introduced. This aspect of the job was possibly the most satisfying, because its aim was to reduce the amount of physical work needed to operate, and also to make it more safe. On the question of cost, obviously it was cheaper to take a job aboard an established scallop diving boat, but on the other hand, if it was desired to invest in a permanent set-up, then that described above was found to be relatively inexpensive.

Although the advantages of working alone were great, it must not be thought that it was an easy existence, because a lot of hard physical work was needed before a sufficient financial return was made. On the other hand, working in this manner proved rewarding and satisfying to many divers. The question of safety was the overriding factor which deterred some divers from working alone, but the record was very good and I know of no one who had any trouble whilst working in this manner. It was up to the diver to form an opinion of how safe his working method was, and if, for instance, he was worried about boats coming and going overhead, then he would work ground where there was little or no traffic. If he still felt unsafe, he would take a man on the boat to tend him, but this meant additional expense and that a certain amount of independence was lost. By far the best way of working was to study the system closely and apply such safety measures as would give peace of mind, as well as successful working.

8 Diving gear

The first requirement of any aspiring scallop diver is, of course, to be able to dive. But after he has learnt and practised the technique of diving he must choose the right diving gear for the method of fishing he intends to adopt.

Diving is prey to many innovations and much time and money can be wasted in determining which are good and which are not so good. Scallop diving certainly does not warrant a fashion display of gear.

This chapter outlines the basic gear needed and the advice it contains is based on practical experience.

Diving suits

Wet suits

Wet suits are available in a wide range, but before buying check these points:

   i Is the neoprene as thick as the manufacturer states?
  ii How easily will the skin of the neoprene tear?
 iii How well are the seams glued and stitched?
 iv Is there enough reinforcing on stress areas?

If you decide to have a wet suit tailor-made, then the following specifications will help. Six and a half millimetre double nylon lined neoprene will give enough insulation and strength to the suit, although the under-vest can be of as little as two millimetres thick. The long-johns should have no zip but just a shoulder fastening. The jacket should be of the raglan shoulder type with hood attached and no zip. For extra warmth it is a good idea to fit a hood to the under-vest as well. The result is a very warm suit. For someone reasonably good with their hands, making up a suit can be very satisfying. Neoprene is a very easy material to work with and, with practice, a perfectly fitting suit can be made. Particular attention should be paid to the stitching and taping of seams, making sure that the thread used will not rot in salt water. A good suit should last about twelve months before the neoprene begins to lose its insulating properties and the seams begin to part. An annual replacement of a suit of this type is not a big expense compared with the cost of replacing other items of diving gear.

Dry suits

Dry suits are used by most scallop divers, in particular the Avon brand. Such items as neck bands, seals and suit inflation valves are common to all suits, so when replacement is necessary these items do not have to be re-purchased. Added refinements to a dry suit are a urinal port, and an automatic pressure release valve. It is a haphazard business to have to hold a cuff open to let the air out of a suit, and it is an improvement to have a valve which does it automatically. Avon also make a heavy duty dry suit of a slightly thicker material, with a much stronger yoke than normal. The only disadvantage is that the suit is slightly more bulky than the standard suit and thus impedes swimming to a certain extent. The secret of getting a long life out of any dry suit lies in the way it is looked after. If possible, it should be washed in fresh water after each day's use and then dried and stowed in a cool place. Exposure to strong sunlight causes the most damage, and care should be taken to ensure that it spends as little time under sunlight as possible. Oil is another thing to be avoided, and any oil on the suit should be quickly removed, using a mild solvent. Both sunlight and oil will quickly cause perishing, and dry suits are not inexpensive items. If looked after properly, a dry suit should last for about eighteen months, but once it shows signs of perishing it is pointless trying to patch it up. Periodic checks should be made around stress areas such as ankles and elbows, to make sure there is no excessive wear likely to cause a bad leak.

The basic dry suit idea has been copied by many firms, the chief difference so far being in the materials used. Beware of the cheaper imitations because, although they may look elegant and professional, some of these suits wear out very quickly. I once bought a new dry suit which seemed entirely suitable for scallop diving. After about three months' wear it began to perish, and shortly afterwards the material split around my ankle. On complaining to the manager of the firm from whom it was bought, I was informed that I was not supposed to do a lot of swimming in it, as that apparently was what had worn it out! Such lessons tend to be very expensive.

Most suit inflation is now done by means of a separate hose off the low-pressure side of the demand valve. This is far better than using a suit inflation bottle, for even when the valve is tight to breathe on it will always let some air into the suit. All these fittings are standard and can be transferred from one suit to another.

Undersuits

Another important point is what to wear under a suit. It is worth spending a little extra on a really good one-piece undersuit. This will give almost complete insulation against the severest cold and yet will allow freedom of movement.

Uni-suits

Another type of suit once commonly worn by scallop divers is the uni-suit. There have been many copies of the original; not all successful. The original is probably about the most expensive of suits on the market, and this is unfortunate because it has out-priced itself with many divers. It is basically a double nylon lined neoprene dry suit with zip entry. However, there can be a problem of water seeping into the suit. The full-length zip is often the culprit, although it is sometimes difficult to pin-point exactly where the water is coming in. Being made of neoprene, it is warm to wear, although it is still necessary to wear a good undersuit beneath it.

Care of diving suits

There are some common-sense ways of keeping a little warmer in both wet and dry suits. Heat loss directly affects the diver's performance on the bottom, so every effort should be made to combat it. Many divers will keep a wet suit long after its insulating properties have declined, just because it looks in good shape. Neoprene can lose its insulation very quickly, especially if worn continually in deep water, so it is advisable to check the condition of the material at regular intervals.

For extra warmth in a wet suit, try experimenting with ways of cutting down the water flow within the suit. Strips of two millemetre unlined neoprene glued around the inside of the jacket sleeve ends and the front of the hood, will form quite an efficient seal and will help cut down the passage of water considerably. A four to five inch wide neoprene strip glued into a ring and worn around the neck will help cut down the heat loss from the base of the skull, and is particularly effective when wearing a dry suit. This neoprene band can be glued inside the bottom part of the hood as a permanent fixture, if desired.

Compressors

A compressor is the most essential and most expensive piece of equipment required, so think carefully before making a final choice. Most portable compressors have a high output and are reasonably reliable. When buying a compressor, bear the following points in mind:

i   What is its output at its maximum working pressure?
ii   In how many stages is the air compressed?
iii   Is the oil circulation within the compressor by pump or by splash?
iv   What is the type and size of the drive used?
v   Is the machine finished well?
vi   How good is the back-up service and availability of spares?

An output of five to seven cubic feet a minute, at three thousand pounds per square inch, should meet the needs of most scallop operations. The air is usually pumped over three stages and the positioning of these is important. If they are in the wrong sequence the compressor will suffer from abnormal vibration, and this will eventually lead to cracked pipes and even a cracked frame. Always ask to see the unit running before buying, as this should indicate any possible trouble with vibration.

A splash feed lubricating system within the compressor is common in many models, but for a long, trouble-free life a pump feed is preferable. This assures that every moving part of the compressor gets an adequate supply of oil.

The power unit of any compressor is very important. Many compressors are made with power units that are only just large enough to run them. This is bad policy because the strain of running at maximum output all the time soon wears the engine out. The engine should always have power to spare, even when the compressor is working at its maximum pressure. The best type of power unit is a small diesel engine, which should be trouble-free, and will avoid the need to keep inflammable petrol on board the boat.

Salt air and water are very destructive to most types of metal, so great care should be taken to see that the compressor is protected from them. A coat of paint on a mild steel pipe will last only a short while when exposed to sea air, yet some manufacturers still send their compressors out like this. All pipes should be stove-enamelled if made of mild steel, but best of all they should be manufactured from stainless steel. This will ensure that no corrosion takes place.

Spare parts are always difficult to get and the problem is made worse if one is working in some remote spot. It is best to try and anticipate breakdowns and have such necessary spares at hand as a spare drive belt, gasket set for the engine and valves for the compressor. The best check on any manufacturers' spare parts service is to question someone who already has one of their compressors. With some companies the back-up service is very poor.

If choosing a surface supply unit, in my experience there is one which meets the scallop diver's requirements better than most others on the market, and that is the Hookah. This low-pressure compressor is light, portable and extremely reliable. The fact that its basic cost is much less than a high-pressure unit has attracted many scallop divers to use it instead of working with scuba.

Air bottles (cylinders)

There have been many improvements in air bottle manufacture over the years, all to the benefit of the diver. The trend now seems to be for very high working pressures and some new bottles can be filled at a pressure of up to 4 000 psi. However, not all compressors fill to that pressure, and ones that do are more expensive. In any case, a working pressure of 3 000 psi is sufficient for scallop diving.

If possible, choose bottles which are not too heavy, because once they are made up into twin packs any excess weight becomes an added working burden. Alloy bottles are very popular with scallop divers and these are very hard-wearing and manageable in twin packs. They have a very high test pressure and, if looked after properly, should last a lifetime. The only fault with them is that their pillar valves corrode quickly if not kept constantly smeared with silicone grease. If they are to be put together as twin packs it is best to get an alloy manifold. These cylinders do not rust, and it is said that if one explodes it opens out, as opposed to fragmenting. A good buy, if steel bottles are preferred, are those supplied by Submarine

Products. These have a low working pressure and they are very durable, with a strong external plastic coating that can be replaced inexpensively.

If you prefer the alloy type of bottle, the Luxfer range is good, and reasonably inexpensive. The price varies in the retail outlets that handle them, so shop around before making a purchase. Diving magazines will give a good indication as to what are reasonable prices for any type of gear.

Demand valves

There is a very large range of demand valves on the market and most of them are very good. It is rare to hear of a valve giving trouble, provided it has been looked after properly. The two-stage single hose model is the most popular today and this is a compact and efficient unit. As valves are not items of high capital outlay and are a very important part of the diver's gear, it is a good policy to change them the moment they start to show signs of wear.

Of the many innovations made in demand valves, the most important lies in the option to choose between an upstream and a downstream type. The earlier valves all used to be upstream on both first and second stages, and although most were excellent value for money, they tended to labour somewhat in deep water. Valves with a downstream first and second stage tend to operate more fluently in deeper water (*Fig 22*).

When buying a demand valve look for one that has a high-pressure take-off, and one, if not two, low-pressure take-offs. The high-pressure is required for a contents gauge, and the low-pressure is for suit inflation. The second low-pressure take-off is useful in case a surface supply is ever used in conjunction with the high-pressure bottle.

Masks

Buying a mask is like buying a pair of shoes, in that what may suit one man may not suit another. Some people like

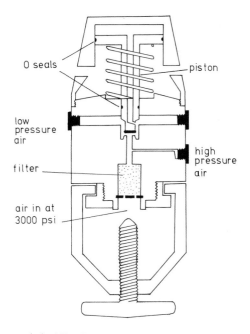

O seals

piston

low
pressure
air

high
pressure
air

filter

air in at
3000 psi

1st STAGE of SINGLE HOSE
DEMAND VALVE

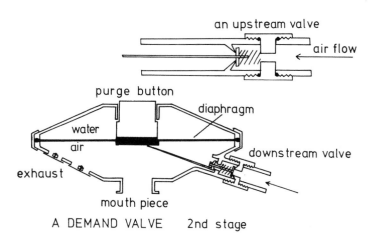

an upstream valve

air flow

purge button

diaphragm

water

air

downstream valve

exhaust

mouth piece

A DEMAND VALVE   2nd stage

Fig 22 First and second stages of a single-hose demand valve, showing
   upstream and downstream mechanisms.

panoramic masks while others prefer the more conservative type; however it is better to stick with one type as much as possible. The full face mask is quite pleasant to wear, as long as you can bear wearing a nose clip all the time. One way around this is to do the same as with hard hats, that is to mount a pad in front of the nose, so that you can block each nostril separately and thus effectively clear your ears. These types of masks seem to be getting more popular, and now you can get one that will take your own single hose demand valve. This unit is quite cheap, especially if you consider that it will take almost any type of demand valve, and it becomes a godsend when there are many jellyfish in the water. The only maintenance a mask needs is a periodic wash in fresh water, and a check to make sure that the seals and the strap are not perishing.

Fins

There are many brands and types of fin available and they differ greatly in performance. Some types of fin are literally not worth wearing, so it is as well to check with other divers before purchase. The Jet Fin brand is very popular with scallop divers and their range is quite large. The original Jet Fin is now quite expensive, and although it is very good, there are others based on the same principle that are equally efficient yet cheaper. Careful attention should be paid to the buckles and the strap, the latter being prone to perish. Sometimes, of course, the rubber in the fin itself will either perish or crack, and this has been known to happen in even the best quality fins.

Weight belts

Most divers make up their own weight belts and there is very little advantage between one kind and another, although it is important to use a stiff belt to support the weights and a sturdy quick-release buckle. A useful modification is the screw-on type of weight. There is, however, a tendency for the weights to fall off if not screwed on tightly.

Harnesses

Again, there is a wide variety available, but one or two stand out as better than others. Until recently most bottles and manifolds were of a standard size, but manufacturers seeking greater variety now provide a multitude of sizes that are not all interchangeable. It is best when buying cylinders and harnesses therefore, to check that they are compatible.

The fibreglass moulded back pack is good and comfortable to wear, but the thickness of the webbing attached to it must be checked, as if it is too thin it tends to twist and become more difficult to put on. The shoulder straps should be easy to adjust, and a strong quick-release buckle is essential on the waist strap. Quick-release harnesses are useful as long as the cylinders are all of standard size. They can be expensive, but as only one is needed to service all the bottles, the cost possibly evens out in the long run. Careful scrutiny should be made of the quick-release mechanism itself, and the quality of the metal used in it. It should be such that adjustment is easy, and the cam action locking the bottle should not need to be forced. Harnesses, if properly made and properly looked after, should last many years. They should be dismantled occasionally and checked for corrosion, but generally should give little trouble.

Life jackets

A great deal of money can be wasted on this item. As its name suggests, this is a jacket that will support a diver on the surface, and as long as it can be inflated easily, there is no need for it to carry any other paraphernalia. They are inflated by means of a small high-pressure cylinder, with the option of mouth inflation, and in fact, when on the surface, the jacket is usually inflated orally in order to conserve the air in the cylinder. There are many good, reasonably priced jackets on the market, but this is an item that needs careful attention and a regular wash in fresh water is essential.

Decompression meters

Decompression meters tend to be very controversial items, yet they are very important for a scallop diver. Curiously enough, most of those who openly criticize the instrument have never worn one, and base their criticism on second-hand information. The main meter in use is produced by a firm called 'SOS', and is strong and reliable. The only fault that can be found with it is that the straps tend to break very easily.

Decompression meters should be checked periodically by lowering two or three into the water at once to a depth of sixty feet, and leaving them there for sixty-five minutes. They should all just be in the red if they are working properly. Their bleed-off time should also be compared, because this is important when using them on a second dive within their six-hour memory period.

9 Home-made gear

For those with practical ability there are many items used in scallop diving that can quite easily be home-made. The following are just a few of the obvious ones, but the building of an air filtration system, or even a complete low-pressure air supply, could well be within the capabilities of a skilful handyman.

Catching bags

Hand-netted

The usual procedure is to make catching bags from old netting. However, by making the netting oneself, the bag can be shaped exactly as one wants it, which can be an advantage. The net-making itself does not take much time, and a complete catching bag can be made in a few hours (*Fig 23*).

Start with a length of two inch diameter rope and splice it into a sixteen inch diameter ring. As a refinement the ends of the splice can be whipped. The netting itself should be made of heavy synthetic twine and some 45 stitches should be started on the rope ring. Although two and a half inches is the standard mesh size for a bag, the size of the mesh can be enlarged or decreased, depending upon how you want the bag to shape up. For instance, the

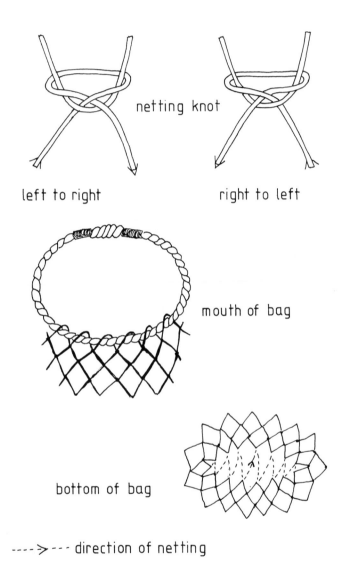

netting knot

left to right   right to left

mouth of bag

bottom of bag

----> --- direction of netting

Fig 23 Making a catching bag; showing how to knot the netting, and details of the mouth and bottom of the bag.

mesh size can be decreased when nearing the end so that the bottom closes in well, and the bottom of the bag can either be finished off with a codend line or can be netted up. When closing it off by netting, proceed as in the diagram – pulling in the stitches as you go.

Making catching bags can become quite addictive, and with practice bags can be made exactly as desired. For example, a bag will be stronger if it has double netting in areas where much abrasion is likely to occur, and this can be done by doubling up the twine on the netting needle.

Using a net panel

This is a satisfactory way of making a catching bag and is quick, although the final shape will not be as practical as the hand-netted one. Two and a half inch mesh is a good size to use, and the heavier the twine the longer the bag will last. The net panel should be cut about 12 meshes deep and 35 to 45 meshes wide, depending on the final shape desired. The panel is netted on to the rope mouth and finally knotted into position as evenly as possible. The side and bottom of the panel is netted together neatly and strongly, and if extra protection is required, a small panel can be stitched around the bottom quite easily (*Fig 24*).

Diving ladder

A diving ladder should be such that it enables the diver to get back on to the boat fully kitted up, and should be secured in such a way that it is not likely to be displaced while someone is on it. It should also be light and easily stowed away when not in use. The one illustrated in *Fig 25* will easily fit on to the gunwale of the boat, and once the ladder is in place it should be secured with a safety rope to prevent it from being accidentally knocked out of position. Two inch diameter steel tube is adequate for the central stay of the ladder, and the rungs should be of three quarter inch steel rod. The actual length of the ladder will depend upon the height of the boat's gunwale from the water.

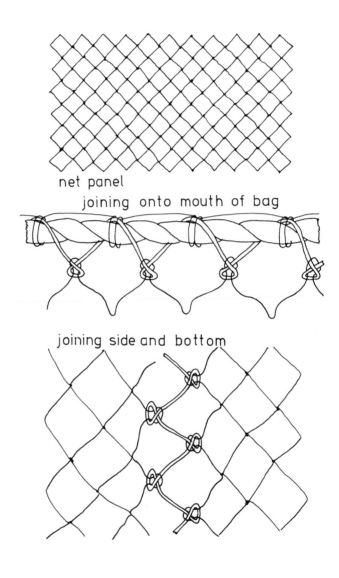

net panel

joining onto mouth of bag

joining side and bottom

Fig 24 A catching bag made from a netting panel, showing how the bag is attached at the mouth and joined at side and bottom.

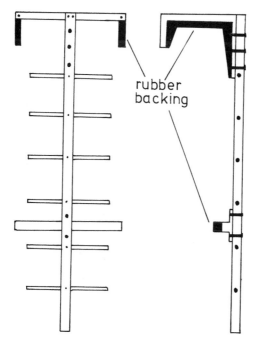

rubber
backing

Fig 25 A simply-made diving ladder to hook on the gunwale of the
diving boat.

Keep bags

Two or three dozen keep bags are essential if the scallops
ever have to be put back in the water, and they are easy to
make (*Fig 26*). If plenty of strong second-hand netting can
be acquired, the only other item needed is rope of about
three-eighths to half an inch diameter (8 to 12 mm) for
drawstrings. The smaller the mesh size, the more effective
the keep bag, but it is not always possible to get small-
mesh netting, because it is in much demand for covering
creels.

The shape of a keep bag is not as critical as that of a
catching bag, so when the panels are being cut out of old
netting, they can be whatever shape is possible from the
net available. The easiest way is to follow straight lines in
the netting and cut a rectangle that can be folded in half

and stitched along side and bottom. A draw-string can then be woven into the mouth of the bag and spliced to form a loop. If there is enough net available to cut out panels in the shape of a catching bag then this is pre-

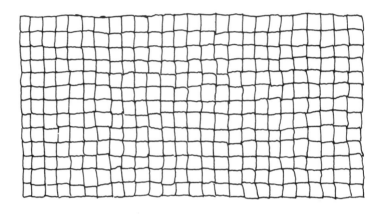

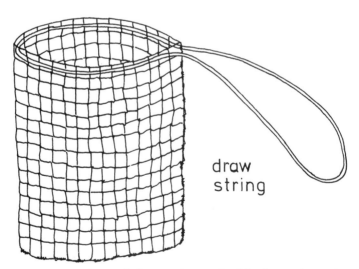

draw
string

Fig 26 A keep bag made from netting panel, with a draw-string to close the mouth.

ferable. Once full, this type of bag will be seen to 'give' more, and thus allow the enclosed scallops more freedom of movement.

Ropes and lines

It is usual nowadays to use synthetic rope for most purposes, and it should be remembered that this kind of rope floats. The result of this is that when a few bags of scallops are 'shot away', all the slack line will float around the marker buoy. This floating rope is a nuisance to any boats passing nearby, and can lead to bags being lost if a line is cut by a propeller. The simplest solution is to tie a small weight like a shackle on to the line, some five to ten fathoms from the marker buoy, and vary the position of the weight on the line according to the depth of water.

When making up any type of line, metal snap-on clips prove very useful, especially when working alone. These clips are inexpensive yet strong enough to cope with most of the problems the scallop diver encounters. They can be attached to the main line by a snood, in order to make 'shooting the catch away' a simple process, or can be used on the diver's surface line to ensure easy release from the catching bag.

Most popular ropes today are synthetic and these have different characteristics from the older type made of natural fibre. These characteristics must be borne in mind when making up lines, whether diver surface lines or otherwise. Generally, a synthetic rope is about twice as strong as a similar-size rope of natural fibre and has the added advantages of being rot-resistant, cheaper, able to 'give' more, and is available in a great many varieties. However, it is not as easy on the hands as natural fibre and has a very low coefficient of friction, which means that care must be taken when joining it. To be on the safe side all knotted joints should be finished off by tucking the short end of the rope into the main line, thus effectively stopping the springing action when the end is left free. If this type of rope is ever spliced, extra tucks should be made to compensate for its tendency to slip.

'Diver down' flag

The most important quality of a 'diver down' flag is that it should be easily spotted by passing boats. It may therefore be better to cut the shape out of a piece of plywood and paint it in the appropriate colours, thus creating a flag that does not lose its meaning when the wind dies down. Whichever way the flag is made, it must always be removed when the diver is out of the water.

If the flag is to be made of cloth then this should be chosen in as light a grade as possible, thus enabling it to keep its shape in the lightest of winds. A minimum size for the flag should be fifteen inches by thirty inches, but if it is to be bigger, bear in mind that it will need a stronger wind to fill it out. Stitching the materials together is a simple process but always use a good strong twine, especially at the free end of the flag, where the action of the wind can be very destructive (*Fig 27*).

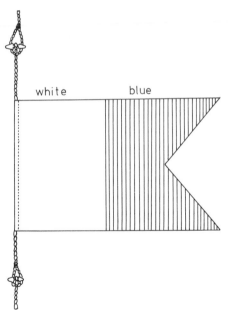

white   blue

Fig 27 A home-made 'diver down' flag.

10 Alternative forms of fishing

It would be wise to buy a boat which would be suitable for an alternative form of fishing. Inshore fishing can be an interesting and rewarding occupation, and even if it does not appeal as much as diving, the fact that the boat is suitable for a different job will make it more saleable. Scallop diving is not as time-consuming as most other forms of fishing, so it is possible to dabble in other methods when the day's dive is finished, and thus learn something about them and possibly make extra cash in the process. However, it must be borne in mind that it is not always possible to carry large amounts of extra gear on the boat, especially if the scallop ground is changed frequently.

A complete change from scallop diving should be viewed and treated with caution, because additional fishing gear will involve extra expense. Scallop diving gives a fairly secure income because a diver will usually find some scallops if he perseveres. On the other hand, fishing for other species can be a luckless job, sometimes with no revenue coming in for weeks on end. This is when some people will tend to give it up. Hence the good sense of dabbling in other forms of fishing, while still diving for scallops.

Given that the boat in question is in sound condition and seaworthy, the only really essential items to add to its inventory are a pot hauler and a graphic echo sounder.

With both of these on board the alternative occupations open to the boat will be numerous. The pot hauler can be either a capstan type or one that actually grips the rope. It can be driven hydraulically or mechanically, whichever is the more suitable. Hydraulic drives are more popular nowadays and are far less trouble than the belt and pulley system taken from the engine, which is the mechanical way of powering a winch or hauler. The graphic echo sounder should be such as to give a good reading at depths in excess of fifty fathoms, and should also be able to give a good picture of the type of bottom that it is sounding out.

There are many types of fishing that a work boat can turn to, but the enormous cost of investing in all the alternatives would prove prohibitive. If the boat is fitted with a winch, trawling would be a possible alternative to scallop diving, but it is not included here because trawlers are usually much larger than the average diving boat. Of the many types of inshore fishing which could be undertaken and could prove interesting and rewarding, because of cost our discussion has to centre around the one or two forms which would probably be the best investment. This will be determined first by the practical application of the boat, and second, by its relevance to the area in which it is intended to work. In order to give some idea of what is available when changing to another form of fishing, I will describe five types of fishing that are popular around the British Isles: lobster potting, crab potting, prawn creeling, 'queenie' trawling and long lining.

Lobster potting

Fishermen tend to be very conservative and sometimes secretive over ways of fishing for lobsters, and it is interesting to look at the many different types of gear that are set around the British Isles in pursuit of this shellfish. The various methods have stayed the same for many years and have usually been very effective for their particular areas.

Lobster fishing is basically seasonal, although some boats try to continue it all the year round. Catches fall during the winter months but this can be compensated by a sharp rise in price. As opposed to the crawfish, the lobster is not migratory and may travel as little as five miles during its whole life. Thus, once a suitable habitat is found, it is unlikely that the lobsters will move away, and an eventual decline in stock will be due to over-fishing. With the large increases in price in winter, it is sometimes advantageous to store some of the catch when plentiful to sell later. With their claws bound, lobsters can be stored in the water inside fish boxes for quite long periods without any solid food. A diver will often come across lobsters that have literally outgrown the mouths of their 'hides', and are thus stuck inside for good. In this position they must live mainly on plankton and their diet is severely restricted.

Diving for lobsters is not easy and is very controversial. Over the last few years it has caused a good deal of trouble with local lobster fishermen. Most of the scallop divers I know have brought up lobsters only for their own consumption, but there have been others who have tried to make a living out of it. Most divers, however, believe this to be unfair competition to the traditional lobster fisherman.

Diving, of course, can be a great advantage when looking for potential lobster grounds, but that is as far as the two forms of fishing should mix. Relations between fishermen and divers over lobster diving have been strained in the past, and there is still the occasional conflict when divers land any large quantity of lobsters. There has then usually been an accusation that the diver has robbed a creel or string of pots.

Equipping a boat with lobster gear can be very expensive, and this expense increases when gear is lost in bad weather. The lobster pot based on the Cornish design (*Fig 28*) is becoming more and more popular, and increased landings have shown it to be superior to the traditional creel. However, the creel was, and is, quite easy to make oneself, which eases the cost of operation and replacement (*Fig 29*). Most lobster pots are manufactured in metal and coated with plastic to protect the metal. They are covered with synthetic netting, and the eyes are of cane and easily detachable. Each pot is quite expensive, and as at least two hundred would be necessary to make it a really profitable proposition, then the capital outlay would be great. In addition to this there is the expense of the rope to rig the 'fleets' of pots and the 'dan buoys' to mark them.

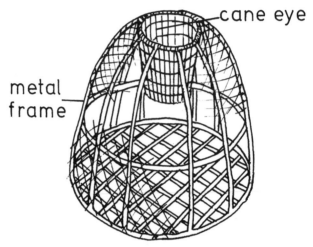

Fig 28 A Cornish lobster or crab pot.

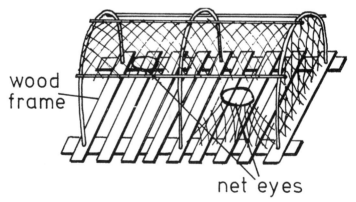

Fig 29 An easily made creel for lobsters or crabs.

The traditional way of fishing for lobsters used to be in either small fleets or single ends (*Fig 30*), in shallow areas of rock bottom. This proved very successful for years but

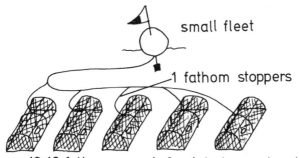

small fleet

1 fathom stoppers

10-12 fathoms apart for lobster and crab

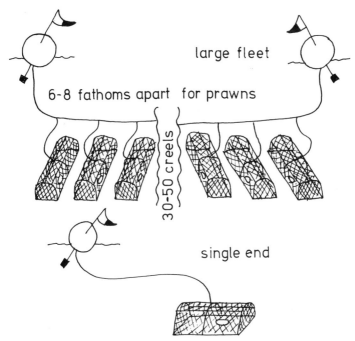

large fleet

6-8 fathoms apart for prawns

30-50 creels

single end

Fig 30 Methods of fishing lobster or crab pots: *top* small fleet; *centre* large fleet; *bottom* single end.

the loss of gear due to bad weather was excessive. Although many lobstermen still set their pots in this way, there has been a recent tendency to fish reefs of up to forty fathoms and over with fleets of forty to fifty pots. At this depth the pots are not readily affected by bad weather, and losses tend to be minimal except for those swept up by trawlers. It has also been found that, contrary to popular belief, lobsters can be quite prolific in some deep water reefs.

As with most types of shellfish, the areas to look for are the edges of rocky outcrops, and especially areas where there is rock debris and weed, because this affords the lobster a better chance of finding himself a suitable 'hide'. Baited with salt herring or mackerel, the 'pots' are fished as close to the rock edge as possible and hauled each day.

Crab potting

Fishing for crabs can be carried out alongside scallop diving because crabs tend to be quite prolific on scallop ground. They can also be worked as an effective alternative to lobsters because the gear involved is much the same. Crabs are more destructive than lobsters, so the pots or creels should be covered with thick, strong netting. If the pots are left down for a number of days without hauling, then it is quite common for the crabs to claw their way through the netting and escape. They are also much more destructive to themselves than are lobsters, and can literally claw each other to bits in a confined space. This is a big problem when boxing them for market, because for at least five minutes, until they settle down, they fight each other with great venom.

Diving for crabs could be a viable proposition, if it were not for their practice of attacking each other. I have found that the only way of overcoming this is to pick up crabs along with scallops, or even sea urchins, the one effectively keeping the others apart. One other problem is in the use of net catching bags, because it is almost impossible to part a crab from a piece of netting once it has got a good hold. Strong polythene bags are the answer to this, but they offer more drag in the water than net bags. Fishermen do not seem to object to divers picking up crabs because of the relatively low price per shell, and the migratory nature of the crab itself. It would thus be unlikely that a diver could actually clean out an area.

Diving is a positive advantage when fishing for crabs because good ground is often spotted, and the diver need only reflect back for a moment to remember where there are abundant supplies of crabs.

Unlike lobsters, crabs will not stay alive for long after being manhandled. If floated in fish boxes for any length of time they will die, so most fishermen try to land them the day they are caught. Keeping them for any length of time is not so important anyway, because the price does not fluctuate much over the year. The latter part of the summer is the peak in most areas for crabs, and they start to decline around Christmas or early in the New Year. It is therefore feasible to change from lobsters to crabs.

On the whole, good crab ground is found at depths of ten to twenty-five fathoms, depending on the time of year. In cold weather they tend to lie deeper, and although rock edges seem to be popular with them, they will also be found on more open areas. They dig themselves deep pits on the sea bottom if in an area for any length of time and, as the water gets colder, they let the sand fall in around them until only their backs are exposed. When this happens, they will rarely be attracted to a baited pot.

If the pots are 'shot' on to the rock itself then few edible crabs will be caught. They are usually 'shot' a good distance from the edge at first, and separate pots are put out at various depths to determine at which depth the crabs are most abundant at one particular time. The edge in this case is only used as a bearing for placing the pots and it is not important to move in close to it. Crabs prefer fresh bait and plenty of it, and it is important that it be tough so that it will not be ripped apart too easily.

Lobster pots and creels are equally good for catching crabs, but the pot is becoming more popular than the creel. The major disadvantage of the pot is that the crabs have to be taken out singly, which is time-consuming. On the other hand, the crab finds it difficult to escape from a pot. Emptying a creel is much easier because a door can be built on to the end, which gives quicker access. There have been many innovations to the creel in crab fishing, and the Norfolk creel is a good example of this. In the Norfolk area they use a tunnel, as opposed to an eye, and this is very effective in stopping the crab escaping. In some areas, very large creels are made, with as many as a dozen eyes in them. These are heavily baited and lifted every three or four days.

On the whole, less expertise is needed for fishing crabs than lobsters, and the job is more routine and boring. However, it can give good returns and at least does not require any extra capital outlay, as long as the boat already has lobster gear aboard.

Prawn creeling

This type of fishing is not as common as lobster and crab fishing around the British Isles, but in the areas where it has been practised it has proved profitable. The capital outlay for gear is very high, and none of the creels discussed so far are suitable for this type of fishing. To make the job worthwhile some four hundred creels are needed, and these each cost about half the price of a lobster pot. A large quantity of rope is required because prawns are fished at depths of between twenty and sixty fathoms, depending on the type of bottom, and at this depth diving is of little use. The rope is usually about one and a half inches in circumference, not necessarily leaded, and large buoys are needed to mark each end of a fleet. At this kind of depth it takes too much time to work small fleets of creels, so most of them support thirty or more.

With such a high capital outlay most boats occupied in prawn creeling try to work all the year round, and they only have to stop occasionally because of trawlers fishing in the area. Although the job is not as interesting as lobster potting, it does give satisfaction and can prove quite profitable. It also lends itself to innovations in catching techniques, and the more adventurous fishermen are quick to try them out.

The three types of shellfish mentioned so far are in big demand, both in this country and abroad, so there is little chance of their prices dropping so low that it would not be worthwhile fishing for them. Fishing for shellfish in the ways described is very interesting and can also be enjoyable, especially for those fishermen who like to experiment with different catching techniques.

'Queenie' trawling

Trawling for 'queenies' with a beam trawl is well within the capabilities of a small fishing boat. As only one warp is needed to tow a trawl of this nature, a capstan or hydraulic line hauler is suitable for hauling purposes.

'Queenies' are of the same family as the scallop but their habits are much stranger, and to fish for them properly one needs a good knowledge of their ways on the sea bed. Watching them move around on the bottom is somewhat like watching butterflies on a flower bed, but in the case of 'queenies' there seems to be no organization in the direction in which they are swimming. They lie in very dense patches on a coarse mud bottom, and the really thick patches tend to be around the rock edges. There have been many stories about their apparent mass migrations, but I have yet to see beds of them actually disappear.

'Queenies' tend to lie in deeper water than do scallops, so diving in search of beds is not a viable proposition. However, some quite good spots are often found in shallower water, and it can usually be assumed that the same bed runs deep as well, which can be quickly proved with a test drag. The biggest problem is to get a good stretch of clear ground on which to operate the trawl, and although some of the shallow water beds look promising, they are usually strewn with rock and weed. Diving does have some use in this job, especially if the trawl is not operating properly. It is feasible for the diver to actually watch a test run in shallow water, and thus see whether modifications are needed.

Beam trawls can easily be made up by the fisherman, as long as he has access to a welding set. A useful size of beam to be towed by a thirty foot boat is from six to eight feet. The codend of an old trawl will make a useful bag for a beam and these can usually be found at any busy fishing port. To protect the bottom of the trawl from being worn away on the sea bed, an extra piece of netting or even a sheet of canvas is worked into it. Chain is necessary for the mouth of the trawl, and a further five or six fathoms will be used as a leader to the warp, to keep the mouth in constant contact with the sea bed (*Fig 31*).

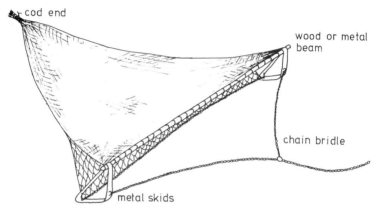

Fig 31 A beam trawl for 'queenies'.

Obviously, a good graphic sounder plays an important part when trawling for 'queenies', because it is all too easy to get the trawl snagged on an outcrop of rock if it is not noticed well in advance. The net codend is easily torn on rock, and 'queenie' fishermen must be prepared to put in many hours of mending if they are to operate efficiently. The common practice is to tow the trawl for anything from twenty to forty minutes before hauling it, and by a system of trial and error the fishermen determine the approximate position of the thickest patches.

'Queenie' trawling can be a little boring if kept up constantly, so many fishermen combine it with some other form of fishing, perhaps on alternate days. With the trawl being relatively cheap, 'queenie' trawling is a good alternative to scallop diving. The market for them is very good, and if landings are of a reasonable quantity, most processors will send a truck to remote places to collect them.

Longlining

Most fishermen who have had experience in longlining will agree that it gave them great enjoyment, although it involved a lot of hard work. The advantage of this method of fishing is that nearly all the fish landed are of very good quality, because it is only the fittest of fish that go for the hook. This means that the best prices are obtained for those landed. It does, however, demand much expertise and patience, but in the long run the rewards are fairly good. The gear itself is relatively inexpensive and easy to set up, and repairs are cheap and easily done. In recent years many innovations have been made in the art of longlining, and these have helped to take much of the monotony out of the job. A machine can now be purchased that baits the line and shoots it away, as well as hauling it in, taking the fish off the hooks and recoiling the line. In this case, the fisherman has plenty of time to decide where to shoot the line, and to take pains to see that everything is functioning properly. As yet, this kind of equipment is well beyond the reach of the average inshore fisherman's pocket, so in giving a general description of longlining, I shall stick to traditional methods.

The lines themselves are made up into baskets of about two hundred hooks each, and these are spaced at one to two fathom intervals. The hooks are attached to the main line by what are known as snoods, and rings of cork attached to the tops of the baskets support the hooks, and prevent them from becoming tangled. When baited, the hooks are left lying over the side of the basket in the order in which they were coiled in the first place, and this

101

ensures that they leave the basket smoothly when being shot away. The lines can be joined together to carry a thousand hooks or more if required. An anchor is attached to each end to secure the line to the sea bed (*Fig 32*). There is a pick-up rope and buoy at either end of the line, and sometimes one or two more are placed at intervals along its length, to ensure that parts of the line will be recovered if either of the two main pick-up ropes are lost.

Longlining is usually carried on in quite deep water, where fish such as haddock, cod, *etc*, can be found, so diving is not of much use. However, good patches of plaice and skate are often seen in shallower water while one is searching for scallop beds, and it can often prove worthwhile to shoot a line over them.

On the whole, longlining seems to have declined in importance during the past decade, due to the introduction of more sophisticated methods of catching fish. However, with the emphasis now on fish conservation and the higher prices being paid for good quality fish, longlining is coming back as a good method of satisfying these new demands. It is a method of fishing that lends itself to much experiment in the form of different baits, different sizes of hooks, varied hook

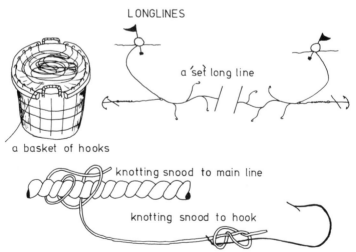

Fig 32 How longlines are assembled and set.

spacings, and ways of shooting away and hauling. Many fishermen find the experimental aspect of the job most satisfying, and it would certainly appeal to those who have a practical nature.

The five types of inshore fishing discussed above will give some idea of the scope a small fishing boat can have, and it must be remembered that there are many more alternatives that could prove equally rewarding. Among these are: ground netting, tangle netting, beam trawling for fish, potting for dog whelks, drift netting, handlining, and so on. Almost all of these will involve some kind of capital outlay, so a boat will possibly be restricted to investing in only two or three of them, but it is a consolation to know that there are so many alternatives available for a well equipped boat.

Although diving does not play a prominent role in the activities discussed, it is more of an advantage than a hindrance. What better aid to catching fish is there than actually being able to watch them in their own environment, and what could be more satisfying than to realise that part of your success is due to this ability. It is interesting to note the number of divers who actually end up participating in some form of inshore fishing once they have given up scallop diving. It seems to be a very attractive activity.

11 Abalone and crawfish diving
– a comparison

An interesting comparison can be made between scallop diving and the longer established abalone and crawfish diving industries. Valuable techniques have been developed by abalone and crawfish divers, and both types of fishing are still changing. As with scallops, these changes are due to fluctuations in price, and the fact that both abalone and crawfish are becoming scarce. When prices are high and catches plentiful it is always worthwhile fishing in a big way, with as many divers as possible, but the burden of running a large outfit becomes apparent when catches fall, and especially when there is a drop in price. Then the bigger boats are replaced by small one-man units, which are cheaper and less worrying to run.

Abalone diving has undergone more changes than crawfish diving.

Regulation of catches

The biggest single influence on both types of fishing has been the introduction of licences in most of the heavily fished areas of the world, and one wonders when this kind of forward thinking will reach the scallop diving industry!

The introduction of licences in such places as Australia, New Zealand and California has certainly helped to conserve stocks of abalone and crawfish, and it is interesting to note the effects that different types of licence have had in individual areas. In Australia, for instance, each state, except New South Wales, introduced its own individual licencing and conservation scheme, and all proved in their own way to be very effective. New South Wales, on the other hand, has been left with no abalone industry of any significance, as it took only five or six years to clear the coast.

One very effective form of conservation was introduced in New Zealand to limit the landings by divers of crawfish, and that was to ban the use of compressed air diving for that particular type of shellfish. This meant that divers had to revert to snorkelling, which was physically much more taxing.

Most conservation agreements on abalone involved a size limit, and to enforce this the meat had to be landed in the shell. The Mallacoota Divers' Co-operative in Australia introduced a scheme that encouraged divers to select only the best and to leave the smaller ones to grow. The abalone were divided by weight into three grades: A (2 fish to the pound), B (3 – 4 fish to the pound) and C (4 – 8 fish to the pound). A diver in the co-operative was not allowed to have more than 10% of his catch as C grade, otherwise he was fined. Premiums on grade A encouraged divers to be more selective in their fishing. This co-operative became very successful and grew to having its own decompression chamber, its own ice plant and distributing system.

Administration of licences in some areas tended to be very strict while in others they were more lenient. In Tasmania, for instance, the licence belonged to the diving boat, which was allowed in turn to employ a certain number of divers. This discouraged the one-man units and helped to keep a closer check on the divers' activities. Those boat owners who held licences were in an enviable position, and large sums of money became involved in licence exchanges. Some other states in Australia operated a licencing scheme that restricted divers to certain areas of the coast, and they were forbidden to enter any other area not covered by their licence. This was another good way of checking divers' activities. Other states were more lax with their legislation, and allowed licence holders to fish where they pleased.

Abalone diving

Abalone are fished by divers commercially in many parts of the world, and the main areas are Australia, California, New Zealand, South Africa and Chile. Diving has started in some areas in the Channel Isles, but there are disputes over sovereignty in the best ground, so the growth of the industry is delayed.

There are many species of abalone and not all are commercial. Some, like those fished in New Zealand, are caught for their shells, while most of those in California are used both for shell and meat content. Out of the nine species found in Australia, nearly all are fished just for their meat content, as their shells do not reach the same standard as those in New Zealand and California.

Abalone are found in commercial quantities at a depth of anything from one foot to 150 feet of water. Being limpets (gastropods), they are mostly found on solid rock reefs and prefer areas of strong tides and rough, exposed water. When an area is fished properly, the abalone will work their way back quite quickly, if allowed to. On the other hand, if they are over-fished, predators will take over. The sea urchin is a great predator of the abalone, and once they have become predominant in any area there is little chance for the abalone. Certain species of abalone prefer proximity to weed, and once they have found this environment they camouflage themselves, so making it difficult to spot them. Other species will show a preference for one type of rock over another.

In California lately there has been friction between abalone divers and naturalists, because a certain species of otter is steadily denuding some abalone beds. Divers want the otters to be kept down by shooting, but have met with a lot of opposition.

The Australian abalone diving industry is interesting to study because all nationalities and all types of shellfish

106

divers have tended to converge on the productive areas. With the recent introduction of conservation laws governing the divers' activities, the industry has settled down to be stable and profitable, with the prospect of many good years of diving ahead.

The industry started in Australia in the early 1960s, when the beds were very extensive and thick. At that time all the diving was with scuba, and from large diving boats. The meat, at this stage, was taken out of the shell and dried in the sun. As the abalone became scarce, so the big diving boats began to disappear and were replaced by small, fast, one- or two-man units. The second man was quite often taken on just as a sheller. These boats could be transported by road to any part of the coast. The smaller boats were also better at negotiating such things as sand bars, which were troublesome to large boats when sailing to and from grounds.

The transition has now progressed one stage further, to slightly larger, faster boats as described in Chapter 4.

The abalone divers' introduction of low-pressure air, and the benefits it has for one- or two-man diving outfits, have been described in chapter 5.

In Australia there is no more drying of abalone; now it is all canned. The fish is either shelled on the boat or shelled ashore. In Tasmania the abalone are sometimes put into tanks of salt water to keep them alive and fresh, but in most cases, if divers want to keep them fresh for any length of time, they salt them. As the abalone goes off very quickly, the longest it can be kept in a brine of this nature is four to five days.

There are freezers in all parts of Australia, so when the abalone are loaded they are immediately blast frozen and sent on to the canneries. The Victorian Canning Company in Melbourne is responsible for the success of abalone diving in Australia. Started by John Stroke, it has distribution outlets all over Asia, and has helped to keep a stable price for abalone meat. It must be remembered that in places like Australia, with its great distances and extremes in temperature, working with perishable commodities can be a gamble, so the establishment of reliable and efficient processing plants and distribution networks is very important.

Crawfish diving

Crawfish, like abalone, come in a variety of species and are found in many parts of the world. They have probably been the fish most sought after by shellfish divers and their catching has caused the most controversy. All the comments about the unsafe way in which shellfish divers work, stem from the heyday of crawfish diving, as does the friction between divers and fishermen. It would be foolish to say that crawfish diving is a safe job, in fact of all the types of diving discussed so far it is probably the most dangerous. However, those people involved in it are certainly not irresponsible and do not deserve the name of 'cowboys', with which they have been so often branded. The problem with crawfish diving started when catches began to dwindle and the divers had to fish deeper to earn a good living; as the price per fish was very good, it was worthwhile to do so. With scallops and abalone the price per shell is relatively small in comparison, so working deep is not always profitable because of the greatly reduced diving time.

Crawfish like hard ground and tend to congregate in small communities. They are migratory to a certain extent, and one cannot be sure that they will be in the same spot today as they were yesterday. Observation shows that different species have their own areas, determined by temperature.

The three species of crawfish found in Australia are in distinct areas, overlapping very little. The red craw (averaging six to eight pounds) and the green craw (averaging one to two pounds) are the most commercial, and are both potted as well as dived for. The red craw is the one mainly found around the coasts of the British Isles. The coral craw (averaging one to three pounds) is a vegetarian, and is therefore not potted. Red crawfish tend to favour deep holes and are harder to catch by diving than green crawfish, which are very sociable and live more in the open. It is a general rule in crawfish diving that as soon as one is spotted, one should look for others nearby, as there will be some around somewhere. The octopus is the main predator of the crawfish, and can reduce whole communities to almost none.

When crawfish diving first started it was a bonanza for many divers. Areas soon became heavily over-fished, and friction between fishermen potting for them and divers diving for them came to a head. Legislation was introduced in many areas to try to conserve stocks and those areas that failed to do this, as was the case on the Cornish coast, soon became practically fished out. Both New Zealand and Australia banned the use of compressed air in diving for crawfish; the Channel Islands introduced licensing, while the Irish Republic banned all types of shellfish diving. These moves have at least appeased the local fishermen in a small way, but for quite a time after the various legislative measures were introduced, divers were using crawfish pots and filling them while they were on the bottom.

Large boats with three to five divers aboard have been commonplace in crawfish diving, and there has been little deviation from this as the years have passed. One-man units are not as common in this industry as in abalone or scallop diving, due mainly to the fact that there have been few entering it since the bonanza ended. Those who have stayed in the industry and have ridden out the hard times, now use the same method with which they started out. With working depths of 120 to 200 feet, surface supply units are of little use, so scuba sets are mainly used. To work at this depth big sets are needed, usually ranging from twin 'seventies' to twin 'eighties'. The correct decompression is vital when working with this amount of air at such depths, and this is where some crawfish divers have become negligent. Divers regularly go on into 'stop-time', or do repetitive dives, which is against the recommendations of the diving tables. The divers involved are convinced that they can stretch the tables, due to their building up a slight immunity to decompression sickness.

This may be so but it is not a practice to be recommended or condoned.

Strong polythene bags are generally used as catching bags for crawfish, and they are floated to the surface when full, the diver usually surfacing with them. Surface lines are optional and in many cases not used at all. Before the introduction of the floater and parachute, the diver swam the bag to the surface himself.

Evolution of the industries

Abalone diving, crawfish diving, and scallop diving have all been confronted with the same problems during the years they have been in operation, and have all gone through various stages. It is interesting to note the similarity between scallop diving and abalone diving, in that big diving boats were first on the scene, and then the industry settled down to a mixture of smaller, two- and three-man outfits. The Hookah system of surface supply is slowly gaining popularity in some areas. It has many advantages over air bottles; the main ones being: as much air on demand as the diver requires, not having to transport bottles to a high pressure compressor, the ease with which decompression stops can be made.

There have been a few small units diving for crawfish, but generally the larger boat has predominated. It is not impossible to work crawfish from a small boat, and in many ways the practice has the same advantages as with abalone and scallops. However, the surface supply system does not allow the diver to work as deep as he would like; and much of the crawfish ground is very exposed, which is unsuitable for a small boat. Some of the divers working at crawfish part-time use the smaller unit, but these men can afford to miss the days of bad weather, so their small boats are adequate. Divers in the job full-time cannot be so selective about sea conditions and have to take advantage of every available opportunity and this is why the larger diving boat predominates.

The use of buoyancy aids in diving for all three types of shellfish was late in catching on. In the early days of abalone and crawfish diving, the diver used to swim his catch to the surface, which was tiring and unnecessary. The scallop divers at least had their bags hauled for them by the men on the surface, but they still had the problem of carrying the full bags around on the bottom. The introduction of the parachute to the abalone trade offered a distinct advantage, and the scallop divers eventually turned to a floater system themselves, favouring a rigid buoyancy aid to the flexible parachute. Crawfish divers also adopted floaters and soon discovered the advantage of working with this method.

It will be interesting to see if scallop diving becomes as popular and follows the same lines as abalone diving and, to a limited extent, crawfish diving. Out of the three types abalone diving is the most well-established, but it is not beyond probability that scallops could progress in a similar way. There is no doubt that scallop diving is here to stay, but at present it is suffering from lack of official interest. A co-operative somewhat like the one mentioned in Australia, and some kind of licensing and conservation officially introduced by the government, would do much to assist scallop diving in becoming a recognized and well-run industry. Both abalone and scallops have an advantage over crawfish in that they are more plentiful and multiply at a faster rate. This means that ground is replenished quickly, as long as it is properly treated, and it also means that the diver can always be sure of getting some kind of catch, no matter how small. Crawfish, on the other hand, maintain a high price per fish, so every one that is found is lifted. This has meant that in many areas the fish has become very scarce. Crawfish diving will, of course, go on for as long as there are divers able to go into the water, but it is unlikely that it will become the well-organized and respected industry that abalone diving has become, and that scallop diving is becoming.

12 Diving and scallop farming

One of the most exciting aspects of diving lies in the possibility of cultivation of shellfish on the sea bed. Fish farming offers the shellfish diver a chance to make use of his first-hand observation of marine life in its own environment, which gives him a head-start in an expanding field. Much can be learned by experimenting with caged scallops, and this is something the diver can do in his spare time, with the aim of possibly creating something more viable in the future. After all, there is nothing to be lost by conducting experiments of this nature, for they are neither too time-consuming nor too expensive to set up, and it could be a good means of establishing a profitable scallop farm, without having to worry about the consequences of failure.

The most simple form of scallop farm, and one that is very common, is created by all the divers on a particular boat gathering all the scallops they come across, and then stocking a particular piece of sea bed with those scallops which are below the marketable size. The first problem to arise here is the likelihood that other divers may find this spot and subsequently fish it. This, of course, can be a problem with all types of sea farms, because it is difficult to legislate against divers working where they please. If, however, the 'farm' is made more structural and permanent, then steps can be taken to secure its stocks.

Sites

The first prerequisite of any farm is a suitable authorized site, *ie*, sheltered from the prevailing wind, with a good tidal flow and a suitable bottom. An additional point to check is that there is no build-up of fresh water on the surface, or, for that matter, on the bottom. It is therefore best to choose a site well away from any river mouth. With regard to supplying good feed for the scallops, plankton will be seen to concentrate closer to the surface than to the bottom, so it is wise to raise the level of the farm to be within range of the largest concentrations. Care must be taken at this point to ensure that the cages are not so high as to be affected by fresh rain water lying on the surface.

Cages and enclosures

It has been my experience that caging large numbers of scallops directly on the sea bed is not very successful (*Fig 33*). Although their mortality rate is fairly low, their growth rate is very slow and their quality stays poor. I can only assume that there was not enough food for such a high density of scallops kept in this way. The answer to this

Fig 33 Cages on the sea-bed. Scallops kept in such cages tend to grow slowly and be of poor quality.

113

problem is to raise the scallops in layers, in order to benefit from the denser concentrations of plankton (*Fig 34*). This, of course, adds to the work of setting up the farm, but it helps to overcome the losses of scallops due to predators. Any cage placed on the sea bed, no matter how small the mesh size, will soon become a haven for small crabs, starfish and whelks. It is interesting to note that many of these marine creatures enter the cages when small enough to pass through the mesh, and then grow to such a size that they can no longer get out. Cages lying on the bottom will attract all kinds of sea life, and although not very productive with regard to scallop growth, they are very interesting to observe.

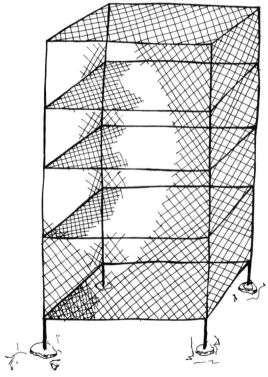

Fig 34 Tiered cages off the sea-bed, in which scallops grow satisfactorily and where they are better protected from predators.

Closing off an area of sea-bed is a good way of starting a rudimentary type of farm, and will at least give a start in building up stocks for future expansion. An area of sea bed, 30 feet square will require 120 feet of 3 feet high net, and this can either be made by hand or out of 'used' net panels. If made by hand, the mesh size can be anything from 6 to 8 inches, so the time involved in manufacture is not long. Once the net is made, its sole rope has to be weighted and its head rope made buoyant. It can then be stood up on the sea bed in the shape of a circle or a square, whichever is easier. A total of eight upright posts should be enough to support a net of this size, and these can be metal stakes set in concrete blocks of about 1 cubic foot. These blocks can then be 'jetted' into the sand to make a permanent and strong support (*Fig 35*).

Once the net is placed around the uprights, the sole rope can be made fast to the bottom by piling rocks on top of it. Although the net's large mesh size will let a certain number of scallops escape, and also let predators in, it will soon become an effective barrier when weed starts to grow on it. The only problem at this stage is caused by the weight of the weed putting strain on the net and causing 'drag' in areas of strong tide. It is therefore necessary to ensure that the uprights are always secure in the sea bed, and that the net's head rope is always taut.

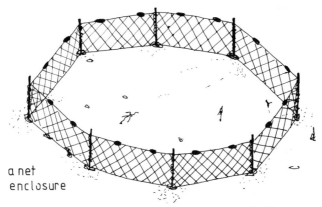

a net
enclosure

Fig 35 A sea-bed enclosure in which scallops can be held temporarily.

115

An even simpler type of enclosure can be made by netting off a small bay or inlet (*Fig 36*). This can be very effective as long as the conditions are right, but suitable inlets are difficult to find. One very important point to watch out for with this type of enclosure is that there is a suitable tide flow. If all the conditions are favourable, a netted-off bay or inlet can give a large area of sea-bed on which to stock small scallops, and should prove very easy to maintain.

With a sea bed enclosure, undersized scallops can be emptied into it from the surface, and stocks of shell can be built up until more effective 'pens' are constructed. By placing a mooring block and surface line in the middle of the enclosure, a boat can get an exact 'fix' as to where to drop the scallops after each day's fishing, and all that is then required are periodic inspections to gather up those scallops that did not drop into the enclosure, or those that escaped. Predators will, of course, be a problem on a farm of this nature, so during the inspection dives they can be cleared out. Starfish will probably be the most numerous predators, and fortunately these can be picked up and disposed of easily. Once gathered, however, they should be taken well away from the enclosure before release, otherwise they will quickly make their way back to it.

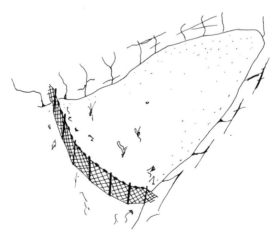

Fig 36 An enclosure made by netting an inlet.

Rearing scallops from larvae

Having created such an environment, it will be noticed that scallops in the vicinity will make their way towards the cages. When it is spawning time this concentration of shell will emit a great number of larvae, and these could provide future stocks if captured. After a few days of free swimming, the larvae attach themselves to sea weed by their byssal thread. This position is maintained for three to four weeks, after which time they are large enough to seek their own position on the sea-bed. By allowing the growth of seaweed in one area, or by even scattering lengths of old natural fibre rope around on the bottom, these larvae can be encouraged to settle.

The collection of young scallop larvae can be carried out quite successfully, and so long as they can be safely 'penned', this brings our scallop farm into the realms of a well-organized unit. It must be remembered, though, that at this stage losses of shell are great, and continual manhandling does not help the situation. It may be considered worthwhile to use the small scallops and larvae merely to re-stock areas of the sea bed that could be dived on at a later stage, but again the problem arises of other divers getting the benefit of one's hard work. It may therefore be a good idea for all the divers in a particular area to jointly re-stock the ground in this way.

Onion bags are the most common receptacle for catching young scallop larvae. The fine net bag is filled with anything to which the larvae might attach themselves, such as rope, old shells or bracken, and then it is tied off at the neck. Strings of these onion bags can be spread at varying depths through the water by weighting them unevenly. When in position, the free swimming larvae will swim through the mesh and attach themselves to the debris inside, but by the time their byssal threads are severed, their shells are too large to slip through the meshes of the bag. In this way the young scallops can be effectively gathered.

The mortality rate between the scallop spawning and the shell being strong enough to seek its place on the sea bed is very high. Although scallops bear both male and female reproductive organs they cannot fertilize their own eggs. Once one scallop starts to spawn, others lying in the same area will also start, but it is still a very 'hit and miss' affair. Laboratory experiments have been carried out to try to monitor this situation, but as yet with little success. However, much good work has been completed by Dan Minchin, a biologist working in Eire in his more natural open-air laboratory.

In his mobile laboratory, scallops were caused to spawn by employing the technique of 'thermal shock'. Those shells ready to spawn were laid out on flaked ice before being placed into the warm water of the spawning trays. The eggs and sperm were mixed together in plastic bins and then poured into a children's paddling pool, moored in the same area of sea that the young scallops were eventually destined for. The larvae were held in the pool for three to four days before being put into the sea, and by this time the small scallops had grown past the stage where they were usually susceptible to a high mortality rate. If they had been kept in the pool for a longer period they would have become prey to destructive bacteria, so the length of this period of unnatural captivity is critical.

These experiments were quite successful, and by using albino scallops (pure white shells) as parents, the effects of settlement on the sea-bed could easily be monitored. They also pointed out just how much more successful it is to use natural rather than man-made environments in conducting experiments.

For the part-time farmer, collection of larvae is the point at which the operation becomes more involved. Whereas rough cages can be constructed to stock undersized scallops, the small larvae can only be restrained by mesh the size of an onion sack, and this calls for more complicated and expensive construction. These shells will have to be kept for some two or three years before becoming marketable, and will have to be graded and thinned out regularly. Placing them on the sea-bed to grow is a good alternative, but the full benefits of the preliminary work would not be felt by doing this. It is therefore preferable to build large cages for these shells, and these can either be designed to be close to the surface or on the sea-bed. However, with the ability to dive, the latter will be found the easier to manage. Cages moored on the surface are susceptible to wave and wind action, and can quickly break up, or even sever their moorings, if not checked and serviced regularly. By building on the sea-bed the need for complicated moorings is overcome, and the result is unobtrusive and stable.

The size, shape and composition of underwater cages can be experimented with over a period of time, or as materials become available. Information can be gathered from observing the constructions in existing farms, but it must be remembered that for the part-time farmer some of the manufactured cages could prove costly. As a general rule, the frames of any cage should be of galvanized steel, or at least plastic-coated, and the netting should be synthetic and therefore rot-proof.

When working from the sea-bed I found it successful to cage my scallops in layers. The first layer was two feet off the bottom, with the successive layers being one foot apart. A cage with floor area 5 × 2ft, in four layers, gave a total forty square feet. There can be many variations with this style of cage, depending on the ingenuity of the builder. Collapsible 'chinese lantern' style cages are not difficult to construct and, more important, are easily handled on the sea-bed. With any size and style of cage there must be a door in each layer at least large enough to allow access of a hand and arm.

The whole business of building cages for stocking scallops can be very absorbing. If your budget is small then improvisations will have to be made with the materials available, and the end result can sometimes be most ingenious. The objective is to create an environment of cages and sea-bed well stocked with scallops, and from this stage interest will gain momentum and expansion will become inevitable.

Running a scallop farm from the sea-bed can be an exhilarating experiment, and the chances are that it will eventually lead to something quite profitable. It takes time and patience to get the whole thing started, but at least it does not require a large capital outlay, so there is little to lose. Farms with cages floating close to the surface, although quite successful, present far more problems than do those run from the sea-bed.

13 The future

The growth of the scallop diving industry has been spectacular over the past few years, and in addition to its role in fish conservation it can be credited with creating employment through processing and handling of scallops, and helping the balance of payments through foreign earnings on exports. It should therefore be encouraged to grow in a well-organized manner, but being so different from traditional forms of fishing it has taken time to become accepted. Methods of working scallops are now becoming fairly standard, and the divers involved are putting a great deal of thought into running cost-conscious and profitable operations. However, if the industry is to expand and take its place in the fishing and diving world, a few problems will have to be resolved. The two most important points for consideration are the relationship between diver and traditional fisherman, and the conflict between the shellfish diver and the commercial diver.

In the past there has been a great deal of conflict between divers and traditional fishermen. The main cause of this controversy was arguments over lobster and crawfish landings, with accusations being made about divers robbing pots. Fortunately, the scallop diving industry has transcended this state of conflict and can now boast a reasonably good relationship with traditional fishermen. This has taken a long time to develop and has been built up mainly by divers servicing the fishing industry, by way of laying and checking moorings, clearing propellers, retrieving lost gear, and minor hull repairs from below the water line. The problem of divers being accepted as fishermen by those already in the business is something that will have to be worked on in the future. At present the traditional fisherman is still sceptical of diving as a legitimate form of fishing, so it is up to scallop divers to remedy this situation by working in a professional and diligent manner. Nothing can gain more respect from fishermen than the sight of hard and conscientious work.

Divers tend to work from many varieties of craft, although on the whole the traditional style of fishing boat predominates. In the past, traditional fishermen have viewed some of the more unusual craft with scepticism, and this has been one of the points alienating the diver from the traditional fishing scene. However, traditional types of fishing are themselves having to undergo changes, with the introduction of fast plastic work boats and the fishermen seem to be accepting these.

Raising the status of the scallop diver to a more professional level is something which should be sought in the future. It is up to the individual diver to ensure that he works in a professional manner and abides by any legislation that is introduced. It is only common sense to work in a safe manner, and although some of the new legislation on diving may seem very restrictive, it must be remembered that its objective is to help save lives.

Foolish talk is responsible for many of the derogatory comments levelled at the scallop diver. It is very unprofessional to make light of breaking all regulations and of abusing decompression times, yet there are still a number of divers who revel in this kind of bragging. First impressions count a lot, and when talk like this is overheard, it creates an overall bad picture of scallop diving, even though the majority of divers do not indulge in this kind of stupidity.

A good way for the industry to raise its status is for all those involved to make an effort to learn as much as they can about all aspects of diving, not only those immediately related to their job, but other techniques which may be applied to it in the future. Although scallop divers become very proficient in their underwater environment, the job only requires the very basic diving skills. The acquiring of fishing skill is a diver's prime concern, but I am sure that a more extensive knowledge of diving practice would benefit the industry in the future. Perhaps if the Government-supported diving schools showed more enthusiasm towards the scallop diver and encouraged entry into their courses, the changes needed in the industry would come about.

Conservation is a subject on everybody's mind at present, and it is a particular bone of contention in the fishing industry. It is often said that the fisherman is his own worst enemy with regard to conserving stocks, because his actions are usually motivated by the axiom: 'If I do not take it, somebody else will'.

Upon close examination many methods of fishing show a high degree of wastage by the very nature of their operation. Take for instance, the ordinary otter trawl, which, being non-discriminatory, will gather everything in its path, be it marketable or not. Once inside the trawl, the likelihood of any fish or shellfish surviving, even if put back into the water after sorting, is minimal. All under-sized fish that get in the way of the trawl are gathered, later to be put back into the sea, dead. This is extremely harmful to future fish stocks, but appears to be unavoidable because of the way the modern-day trawl operates.

The scallop dredge is another piece of equipment that causes damage to the sea bed. Modern dredges with retractable teeth are highly efficient at scooping up scallops, but unfortunately they also scoop up everything else that happens to lie in their way. Furthermore, they are responsible for destroying many scallops during the process of fishing, either by pushing them down into the sand or by actually breaking their shells.

The use of the trawl and the dredge can cause untold damage on spawning grounds, but this is inevitable. The fertile areas around rock edges are usually particularly rich in young marine life, and yet these are usually the spots where the trawl and dredge take their best fishings.

The scallop diving industry can take full credit for being completely selective and conservation-conscious in its operation. The diver only picks up off the sea-bed that which is marketable, and during this process he does not damage any other form of marine life. This is not inferring that the scallop dredge is inferior to diving as a means of fishing, because the two forms can work side by side with little conflict and both be viable forms of fishing. However, simply on the basis of conservation, diving is preferable because of its selectivity and its consideration to the sea-bed.

Following the example of the Australian abalone diving industry, some form of licensing would be good for the future of scallop diving. This would have the advantage of conserving stocks of shell for the future, and, equally important, of raising its status to a more professional level. What form the licensing should take is a matter for debate, but lessons could be learned from those licensing systems already in operation. One way is to license a certain number of boats, with regulations on the number of divers each one can carry. This has the advantage of keeping the system simple but, on the other hand, it places a great deal of power in the hands of a few people. Individual licences might be fairer; divers could operate independently, maintaining the freedom which is one of the attractions of the job.

New technology is being introduced to traditional types of fishing, and the results have already been spectacular. Scallop diving could also benefit from some of these innovations, although generally the cost of setting them up is high. Underwater video, operated from the surface, offers one of the most exciting prospects. The cameras used in these units are very sensitive and can pick out objects on the bottom that the human eye would normally miss. It is therefore feasible that beds of scallops could be pin-pointed without a diver entering the water. The amount of time saved in surveying ground in this manner could be enormous, and a very great advantage is that the capital outlay for a video set is within the limits of a profitable scallop diving boat.

124

Underwater tugs have great potential as another possible innovation to scallop diving. They are small, compact and reasonably inexpensive, and if suitably adapted, could be a great help in pulling the diver, or even the catching bag, along the sea bed. This would mean a considerable saving in effort for the diver, and much more ground could be covered during a dive.

The state of the scallop fishing industry at present is very interesting. Many of the large herring boats have been fitted out with dredging equipment as an alternative to their traditional type of fishing, and this has meant increases in scallop landings throughout the British Isles. This has affected the price, but on the whole 'dived' scallops have maintained a good price, because of their superior quality and undamaged state on landing. Dredging is becoming more and more popular and landings are good, but it is very taxing on the boat and involves travelling long distances to find suitable grounds.

A most interesting question in the scallop industry is posed by farming scallops, because eventually prices will be affected by this activity. The Japanese, in the spring of 1978, showed the effect on the industry of exporting a large amount of farmed scallops to Europe. Prices became very unstable and it took many months before they settled again. However, once again the diver was not at too much of a disadvantage, because of the superior quality of his landings.

It is hoped that the scallop diving industry will settle down in the future, and innovations will be gradually introduced to make it even more viable. The introduction of more restrictive diving legislation could be the only thing to hamper its progress, but hopefully this eventuality will not come about. At present, scallop diving, being an alternative to other types of commercial diving, offers a rewarding and interesting occupation, and should be encouraged to develop.

Bibliography

US Navy Diving Manual. Superintendant of Documents, US Government Printing Office, Washington DC

Underwater Swimming (An advanced handbook). Leo Zanelli, Kaye and Ward, 1969

Commercial Oilfield Diving. Zinkowski, Cornell Maritime Press Inc, 1971

BSAC Diving Manual. British Sub Aqua Club, 1978

The Physiology and Medicine of Diving. Bennett and Elliott, Bailliere Tindall, 1975

Marine Biology. H Friedrich, Sidgwick and Jackson, 1970

The Sea Shore. C N Younge, Collins, 1970

Fishermen's Handbook. Capt. W H Perry, Fishing News Books, 1980

Fishing Boats and their Equipment. Dag Pike, Fishing News Books, 1979

Where to learn to dive in the UK

The British Sub Aqua Club
70 Brompton Road, London SW3 1HA. This club has 750 branches in this country, with some 2 500 members here and abroad. They teach a high standard of diving, extensively covering both theory and practice.

Independent Sub Aqua Clubs
There are non-affiliated sub aqua clubs in the UK which also teach a high standard of diving.

Diving Schools
The following schools are a few of the many which cover a wide range of diving practice, from simple sports diving to deep saturation diving. The commercial courses run by these schools are excellent and teach the novice diver the basic skills needed to carry out his trade.

Fort Bovisand
  Underwater Centre
Fort Bovisand
Plymouth
Devon
PL9 0AB

Diver Training School
  (Devon)
The Quayside
Exmouth
Devon

The Underwater
  Training Centre
Inverlochy
Fort William
Inverness-shire
Scotland

Divers Down
  Diving School
The Pier
High Street
Swanage
Dorset BH19 2AM

UK diving regulations

The Factories Acts define the basic rules and regulations affecting shellfish diving in this country. All the following regulations will eventually come under a single document, called *The Health and Safety at Work Act (Diving Regulations)*.

The Factories Act; Diving Operations, Special Regulations 1960 Administered by the Health and Safety Executive and inspected by the Factories Inspectorate, these regulations apply to divers working within the three mile limit in UK waters. The Health and Safety at Work Act 1974 has now superseded these regulations.

The Submarine Pipeline (Diving Operations) Regulations 1975 Administered and inspected by the Department of Energy, these regulations apply to diving on underwater pipelines in UK territorial waters.

The Off-shore Installations (Diving Operations) Regulations 1974 Administered and inspected by the Department of Energy, these regulations apply to diving from off-shore structures, such as oil-rigs.

The Merchant Shipping (Diving Operations) Regulations 1976 Administered and inspected by the Department of Trade, these regulations apply to diving from ships in UK waters, and to British ships anywhere in the world.

Metric conversion table

1 ft = 0·3048 m.

Feet	Metres	Feet	Metres	Feet	Metres	Feet	Metres
1	0·30	26	7·92	51	15·54	76	23·16
2	0·61	27	8·23	52	15·85	77	23·47
3	0·91	28	8·53	53	16·15	78	23·77
4	1·22	29	8·84	54	16·46	79	24·08
5	1·52	30	9·14	55	16·76	80	24·38
6	1·83	31	9·45	56	17·07	81	24·69
7	2·13	32	9·75	57	17·37	82	24·99
8	2·44	33	10·06	58	17·68	83	25·30
9	2·74	34	10·36	59	17·98	84	25·60
10	3·05	35	10·67	60	18·29	85	25·91
11	3·35	36	10·97	61	18·59	86	26·21
12	3·66	37	11·28	62	18·90	87	26·52
13	3·96	38	11·58	63	19·20	88	26·82
14	4·27	39	11·89	64	19·51	89	27·13
15	4·57	40	12·19	65	19·81	90	27·42
16	4·88	41	12·50	66	20·12	91	27·74
17	5·18	42	12·80	67	20·42	92	28·04
18	5·49	43	13·11	68	20·73	93	28·35
19	5·79	44	13·41	69	21·03	94	28·65
20	6·10	45	13·72	70	21·34	95	28·96
21	6·40	46	14·02	71	21·64	96	29·26
22	6·71	47	14·33	72	21·95	97	29·57
23	7·01	48	14·63	73	22·25	98	29·87
24	7·32	49	14·94	74	22·56	99	30·18
25	7·62	50	15·24	75	22·86	100	30·48

Index

131

Other books published by Fishing News Books Ltd, Farnham, Surrey, England

Free catalogue available on request

Advances in aquaculture
Advances in fish science and technology
Aquaculture practices in Taiwan
Atlantic salmon: its future
Better angling with simple science
British freshwater fishes
Commercial fishing methods
Control of fish quality
Culture of bivalve molluscs
Echo sounding and sonar for fishing
The edible crab and its fishery in British waters
Eel capture, culture, processing and marketing
Eel culture
European inland water fish: a multilingual catalogue
FAO catalogue of fishing gear designs
FAO catalogue of small scale fishing gear
FAO investigates ferro-cement fishing craft
Farming the edge of the sea
Fish and shellfish farming in coastal waters
Fish inspection and quality control
Fisheries of Australia
Fisheries oceanography and ecology
Fishermen's handbook
Fishery products
Fishing boats and their equipment
Fishing boats of the world 1
Fishing boats of the world 2
Fishing boats of the world 3
The fishing cadet's handbook
Fishing ports and markets
Fishing with electricity
Fishing with light

Freezing and irradiation of fish
Handbook of trout and salmon diseases
Handy medical guide for seafarers
How to make and set nets
Introduction to fishery by-products
Inshore fishing: its skills, risks, rewards
The lemon sole
A living from lobsters
Marine fisheries ecosystem: its quantitative evaluation and
    management
Marine pollution and sea life
The marketing of shellfish
Mending of fishing nets
Modern deep sea trawling gear
Modern fishing gear of the world 1
Modern fishing gear of the world 2
Modern fishing gear of the world 3
More Scottish fishing craft and their work
Multilingual dictionary of fish and fish products
Navigation primer for fishermen
Netting materials for fishing gear
Pair trawling and pair seining – the technology of two boat fishing
Pelagic and semi-pelagic trawling gear
Planning of aquaculture development – an introductory guide
Power transmission and automation for ships and submersibles
Refrigeration on fishing vessels
Salmon and trout farming in Norway
Salmon fisheries of Scotland
Seafood fishing for amateur and professional
Stability and trim of fishing vessels
Seine fishing – bottom fishing with rope warps and wing trawls
The stern trawler
Study of the sea
Textbook of fish culture: breeding and cultivation of fish
Training fishermen at sea
Trout farming manual
Tuna: distribution and migration
Tuna fishing with pole and line